STRIPLINE
CIRCUIT
DESIGN

MODERN FRONTIERS IN APPLIED SCIENCE

ARTECH HOUSE, INC., 610 Washington Street, Dedham, Massachusetts, 02026.

STRIPLINE
CIRCUIT
DESIGN

HARLAN HOWE, JR

Microwave Associates
Burlington, Mass.

Introduction

Since the invention of the klystron shortly before World War II, microwave component design has moved from the laboratories of a few far-sighted pioneers to production lines turning out components in large volume for both commercial and military applications.

Stripline circuit design, which started in the early 1950's, has evolved from the black magic techniques of razor blades and glue to a more predictable and precise technology. Although many papers and articles have been written as this evolution has taken place, there has been no central reference on stripline practice since the Sanders Tri-plate Manual was published in 1956. I believe that a need exists for a general text on stripline design and practice incorporating more recent methods. This book was written to fill that need.

I have assembled the book to be a practical handbook, not only to aid the component and subsystem engineer, but also to be of use to the systems engineer who specifies microwave components.

It is organized to progress from materials to basic concepts, from simple structures to more complex circuits, and finally to packages containing combinations of these structures.

Chapter 1 discusses the physical properties of materials in combination with the linewidth, line losses, and power handling capability of the construction, which should be basic considerations in the early stages of any design.

Chapter 2 presents several methods of calculating characteristic impedance and describes basic structures and methods for launching from other transmission lines into stripline packages.

Chapter 3 covers direct coupled hybrids, power dividers and directional couplers. Because these circuits do not rely on parallel line coupling fields, the microstrip designer will also find this chapter helpful.

Chapter 4 is devoted to parallel-coupled lines and includes design equations, curves, and tables for edge-coupled, broadside-coupled, and partial-overlap-coupled lines. These physical dimensions can be applied to the coupled-line devices described in Chapters 5, 7, and 8.

Chapter 5 describes parallel-coupled line directional couplers, including single-section, multi-section, symmetric, asymmetric and non-uniform line types.

Chapter 6 covers the design of low-pass and high-pass filters based on lumped element prototypes. One type of high-pass filter and three types of low-pass filter are described along with their response curves and the design techniques to be used for each.

Chapter 7 includes several types of bandpass and bandstop filters which are suited specifically for stripline applications, including side-coupled types, stub filters, directional filters, and spur-line band-reject filters.

Chapter 8 discusses the application of hybrids to mixers, switches, and other circuits.

Chapter 9 presents a variety of construction technques, as well as multi-layer, multi-media, and other packaging and sub-assembly methods.

Also included is an extensive bibliography for those who wish to pursue specific subjects in greater detail.

During the preparation of this book, there were a number of people at Microwave Associates who made significant contributions. Barbara Leaver typed the entire manuscript from my hieroglyphics and tapes, and also managed the interface with the word-processing computer and photo-composer which was used to generate the text. Her contributions and infinite patience are deeply appreciated. Dianne M. Cotton and Sharon Donahue, members of the Microwave Associates Art Department under the direction of Irene Lambert, were responsible for the final generation of the curves, tables, and drawings. Miss Cotton did all of the figures; Miss Donahue did the composer work on equations and lettering. The response curves were generated by means of MACAP (Microwave Associates Circuit Analysis Program) which was written by Harold Stinehelfer, Sr., Director of the Computer Science Department. The text was read for technical content by Kenneth Carr, Vice-President of Engineering, whose comments and suggestions were extremely helpful. Derwin Hyde, of Artech House, was also of great assistance. My wife, Nancy, proofread the final text and also provided support, encouragement and many helpful suggestions over the years it took to prepare the text.

To all of these people I extend my thanks and appreciation for their help and contributions.

Harlan Howe, Jr.
Acton, Massachusetts
December 1973

Contents

CREDITS

Figure 2-3:   S. Cohn, "Problems in Strip Transmission Lines," MTT-3, No. 2, March 1955, Figure 1.

Figure 2-21:   Matthaei, Young and Jones, "Microwave Filters, Impedance Matching Networks and Coupling Structures," McGraw-Hill Book Company, (New York 1964), Figure 5.07-4, page 206.

Figure 2-23:   L. Young, "Design Chart for Quarterwave Stubs," Microwave Journal, Volume 4, No. 5, May 1961, page 92.

Figures 4-3 and 4-4:   S. Cohn, "Shielded Coupled Strip Transmission Lines," MTT-5, October 1955, Figures 5 and 6, pages 32 and 33.

Figure 9-4:   Hughes Aircraft Corporation, Fullerton, California.

Figure 9-7:   Varian Solid State, Beverly, Massachusetts.

CHAPTER

1

Materials

The starting point for any stripline design must rest with the choice of materials to be used in the final fabrication of the circuit. This includes center conductors, dielectrics and case materials. In this book we are dealing with solid dielectric stripline; hence, the available choice of dielectric material can result in a significant restraint to the freedom of the circuit designer.

Stripline materials come in a wide variety of pure materials and mixtures, all in sheet form, usually with copper laminated to one or both sides. A broad selection of thicknesses, both of the basic dielectric material as well as the conducting material, is also available. While most conductors are copper, several manufacturers are now offering other conductors, such as aluminum. Unfortunately, despite the large selection, there is no universally accepted laminate which is "best". Each material has advantages which make it desirable for some specific applications and disadvantages which limit its usefulness in other applications. As a result, the first step to a successful design is a careful appraisal of the electrical and mechanical requirements. On this basis, the various material trade-offs can be evaluated and a choice made which is most appropriate for the specific application. A number of people (including the author) have attempted to choose a single material which had a reasonable combination of desirable characteristics and then build a design library around it. In general, this has not been a successful approach, since, inevitably, requirements occur which strain one or more capabilities of the material resulting in an inferior product. The fact is that the material evaluation and trade-off procedures should be considered an essential part of any new design project. The factors to be considered in choosing an appropriate material can be summed up by the following list.[1]

1

1. Dielectric constant and its variation with frequency and temperature.
2. Dissipation factor and its variation with frequency and temperature.
3. Homogeneity, uniformity and isotropy.
4. Useful temperature range.
5. Dimensional stability with:

    (a) temperature
    (b) processing
    (c) aging
    (d) humidity
    (e) cold flow

6. Resistance to chemicals and water.
7. Physical factors including:

    (a) tensile strength
    (b) structural strength
    (c) impact resistance
    (d) flexibility
    (e) machinability
    (f) thermal conductivity

8. Characteristics when clad including:

    (a) metals available for cladding
    (b) bond strength
    (c) need for adhesives for bonding
    (d) thermal expansion relations
    (e) blister resistance
    (f) sizes available (both thickness and area)

9. Processing restraints

    (a) chemical
    (b) mechanical

A study of the material characteristics is made difficult due to the wide range data presented for the same basic materials by different manufacturers, many of whom obtain their basic resins from the same source. The reasons for this are twofold:

(1) There are subtle variations in the end product caused by lamination processing techniques.

(2) Most manufacturers really do not know what the characteristics of their product are because of the difficulty and cost of testing.

Thus, they base their specification sheet data on a small sample tested by a laboratory, usually not their own, which has not correlated its measurement methods with those of other laboratories measuring the same basic material for other people. This fault does not occur to as great an extent with the mechanical data where recognized standard test methods have been established as it does with electrical parameters. Thus, where a range of data values is presented, it may be as a result of different reporting methods, rather than the tolerance capability of any given manufacturer.

Glass Reinforced Teflon[R] (Woven)

Glass cloth, which has been impregnated with teflon, (sometimes called teflon fiberglass) is one of the most commonly used materials for operation from UHF to X-band. It has a dielectric constant of about 2.55 and a dissipation factor of around 0.002. The manufacturers who supply it have been able to maintain fairly good uniformity from lot to lot, and it is a reasonably homogenic material. One minor drawback is the fact that due to the woven nature of the glass, it suffers from some anisotropy. In most cases this is not serious, but it can cause manufacturing repeatability problems in coupled line circuits, particularly filters at higher frequencies. The dimensional stability of teflon fiberglass is excellent, both with temperature, humidity, and processing. Like all fiber reinforced laminates, it has a slight tendency toward wicking, thus, to protect the circuit from changes due to humidity, the exposed edges of the boards should be protected. Its resistance to chemicals is excellent, with the result that any processing method may be used. Solder fluxes and solvents have virtually no effect on it.

[R] "Teflon" is a registered trademark of Dupont.

Teflon fiberglass is available copper clad in 0.5 ounce, 1.0 ounce, and 2.0 ounce thicknesses. The ounce designation refers to the weight of copper per square yard of material. Translated into thicknesses, 0.5 ounce equals 0.0007", 1.0 ounce equals 0.0014" and 2.0 ounces equals 0.0028".

The bond strength is excellent and most manufacturers do not use adhesives to maintain the bond of the copper foil. Its machinability is also good, although carbide tools are needed because of the abrasive quality of the glass. Tool speeds must be properly controlled to prevent fraying at the edges.

Data Summary for Glass Reinforced Teflon (Woven)

Dielectric constant 2.4 to 2.62 (2.55 nominal)
Dissipation factor 0.0015 to 0.002
Useful temperature range −100°F to +400°F
Tensile strength 16,000 to 20,000 psi
Flexural strength 12,000 to 15,000 psi
Impact resistance 11 to 15 ft. lbs./inch
Flexibility – good
Thermal conductivity 7×10^{-4} cal/sec/cm^2/C°/cm
Coefficient of thermal expansion 1.85×10^{-5}/°C
Ratio of dielectric to copper thermal expansion 1.32
Thickness range available 0.004 to 0.500"

Glass Reinforced Teflon (Microfiber)

Teflon laminates may also be reinforced with glass which is not woven. At this time there is one manufacturer who produces this version of teflon fiberglass. The glass is in the form of short microfibers which are dispersed in the teflon during manufacturing. The result is a material which has the appearance of slippery cardboard. The major virtue of the microfiber teflon fiberglass is that the problem of anisotropy is solved. The glass is in the form of a random distribution rather than a woven pattern; thus, there are almost no changes of electrical properties with the orientation of the material. Although some small changes in properties have been observed when very thin sections are used, this characteristic has also been noted in the woven variety.

Since microfiber teflon fiberglass has slightly less glass in the mixture than the woven type, there are some electrical differences. The dielectric constant is lower, being 2.32 at L-band and 2.4 at X-band. This reduction in glass content also accounts for a reduction in dissipation factor which ranges from 0.0004 to 0.0008 over the same frequency range. The dimensional stability of the product is among the best of all materials. It has little tendency to warp, is impervious to most chemicals and to all of the standard processing methods. It has good humidity and aging characteristics, but has a greater tendency towards cold flow than the woven type. It is available with any of the standard copper foils as well as other materials on special request. Adhesives are not used for bonding in the case of the copper foils; however, it is not known whether they are needed for other metals.

In general, the processing ease, bond strength, and machinability are about the same or better than the woven teflon fiberglass. Carbide tools must be used for drilling or routing, and the material is slightly better for punching provided it is tooled properly. In fact, it is a generally improved material over its woven relative except for its cost, which due to better characteristics, newness on the market, and smaller volume usage, is greater. With time this problem may be eliminated, in which case the microfiber material may become more widely used.

Data Summary For Glass Reinforced Teflon (Microfiber)

Dielectric constant 2.32 to 2.40 (increasing with frequency)
Dissipation factor 0.0004 to 0.0008
Useful temperature range −100°F to +500°F
Tensile strength 5,500 to 7,500 psi
Flexural strength 10,000 to 15,000 psi
Impact Resistance 1.3 to 2.4 ft. lbs./inch
Flexibility − good
Thermal conductivity 1.8 BTU inches/hour/sq.ft.°F
Coefficient thermal expansion 1.6×10^{-5} to 10×10^{-5}
with direction and temperature range (°C)
Ratio of dielectric to copper thermal expansion −
1.14 to 7.14
Thickness range available 0.005 to 0.125"

Polyolefin

Polyolefin is a material that was introduced about twelve years ago. It is similar in appearance to polyethylene, but differs in that it is subjected to an exposure of radiation which cross-links the molecules, thus providing both chemical and mechanical changes which are permanent provided the material is not exposed to additional heavy doses of radiation. While the manufacturers of these "modified" polyolefins have improved the mechanical stability of the materials, the fact remains that while polyolefins are perhaps the best materials from an electrical standpoint, they remain the poorest from a mechanical standpoint. A variety of approaches has been attempted to improve mechanical properties, the most successful of which seems to be the production of a laminate consisting of one or two ounce copper on one side and 1/16" to 1/8" aluminum the other. The purpose of the aluminum is to impart some stability to a material whose temperature range is limited and which suffers from extreme cold flow characteristics.

The dielectric constant of polyolefin is 2.32 ± 0.01 and is one of the most uniform and closely controlled values available. Similarly, its dissipation loss is uniformly low at 0.0003. It is one of the few materials recommended for use at the higher frequencies for solid dielectric stripline.[2] There are no problems whatsoever with either its homogeneity, uniformity, or isotropy. It is, however, limited to a useful temperature range far less than that available from other materials. Its dimensional stability is poor, both with temperature and processing.

Normal polyolefin (copper clad on both sides), warps badly when the copper is removed from one side. In the thin sections, i.e., 0.005" to 0.010", extreme warp can occur under certain processing conditions. Warp characteristics are improved, somewhat, in the case of the heavy aluminum clad materials, but large sheets will still exhibit substantial distortion after processing.

The material is virtually impervious to chemicals, resisting all but very strong acids and organic solvents at elevated temperatures. The machinability of polyolefin is difficult to describe. It is soft enough to be cut with a pocket knife, and thus does not require special tools. It shears well, and can be punched with proper support. Likewise, it can be drilled and routed with ease, although it is difficult to maintain tight tolerances because of creep and cold flow. Its stability is such that no machining process will alter its properties. No adhesives are used for bonding either the standard copper foils or the aluminum back up sheets.

Data Summary For Modified Irradiated Polyolefin

Dielectric constant 2.32
Dissipation factor 0.0003
Useful temperature range −100°F to +180°F
Tensile strength 3,000 psi
Flexural strength 8,000 psi
Impact resistance 3 ft. lbs/inch
Flexibility − very good
Thermal conductivity 12.7 x 10^{-4} calories/second/cm^2/°C/cm
Coefficient of thermal expansion 10.8 x 10^{-5}/°C
Ratio of dielectric to copper thermal expansion 7.7
Thickness range available 0.010 to 0.125"

Cross Linked Polystyrene (Unreinforced)

Pure polystyrene offers several advantages in terms of its electric characteristics and reproducibility. It is a totally homogeneous material, thus offering complete uniformity and isotropy. Its dielectric constant is 2.53 and the typical reported loss factor is 0.00047 which makes it the lowest loss material after polyolefin. Unlike polyolefin, it is very hard brittle material with very little cold flow. Like polyolefin, however, its temperature range is limited to −80°F to +212°F. Its dimensional stability with temperature is reasonably good, as it is with aging and humidity. It is susceptible to processing problems in that it is readily attacked by most of the solvents and resist strippers used in the photo-etching process. As a result, water based resist systems are suggested.

Polystyrene is available clad with copper, as well as with copper and aluminum. Its bond strength is reasonably good, but not as strong as the glass reinforced laminates. There are severe limitations to its use without mechanical support since it has extremely low impact resistance. This poor impact resistance prohibits fabrication methods such as shearing or punching since the material tends to shatter under this sort of treatment.

Data Summary for Polystyrene (Unreinforced)

Dielectric constant 2.53
Dissipation factor 0.00025 to 0.00066
Useful temperature range −80°F to +212°F
Tensile strength 7,700 psi
Flexural strength 11,500 psi
Impact resistance 0.3 ft. lbs./inch
Flexibility − very poor
Thermal conductivity 3.5×10^{-4} calories/second/cm^2/°C/cm
Coefficient of thermal expansion 7×10^{-5}/°C
Ratio of dielectric to copper thermal expansion 5
Thickness range available 1/32" to 1/4"

Glass Reinforced Polystyrene

Many of the mechanical problems of polystyrene, like those of teflon, can be overcome, or at least reduced, by means of loading the basic material with some other material such as glass. Glass loaded polystyrene is similar to the microfiber teflon fiberglass in that the distribution of glass is random, rather than woven. This loading increases the dielectric constant from 2.53 for the basic polystyrene material to 2.62 for the loaded variety. The loss factor also increases to an average of 0.0014 for this material. The process controls for glass reinforced polystyrene are excellent, with the result that it is remarkably homogeneous and uniform, with only a minimum of anisotropy evident in thin sections.

Its useful temperature range is only slightly better than the unreinforced material; however, its impact resistance, while still not exceptional, has increased by more than a factor of two. Its dimensional stability is better, but processing restraints previously mentioned apply as well here. Most solvents will attack it, and punching or shearing are out of the question. The bond strength of the clad material is greater due to the use of glass fiber and, like the unfilled variety, no adhesives are normally used for cladding either the normal copper or aluminum clad sheets.

Data Summary for Glassfilled Cross-Link Polystyrene

Dielectric constant 2.62
Dissipation factor 0.0004 to 0.002
Useful temperature range —100°F to +212°F
Tensile strength 8,700 psi
Flexural strength 12,000 psi
Impact resistance 0.75 ft. lbs./inch
Flexibility – poor
Thermal conductivity 5 x 10^{-4} calories/second/cm^2/°C/cm
Coefficient of thermal expansion 5.7 x 10^{-5}/°C
Ratio of dielectric to copper thermal expansion 4.1
Thickness range available 0.011" to 0.250"

Polyphenelene Oxide (PPO)

PPO is one of the relatively new materials for stripline use. It is a pure thermoplastic material (unreinforced) with many of the properties of the reinforced laminates. Its dielectric constant is 2.55, which makes it compatible for use with designs which were developed for woven teflon fiberglass and vice versa. Its dissipation factor is 0.0016 as a mid-range value, but it rises rapidly in the region of X-band, thus limiting its usefulness at the higher frequencies. Since it is a pure material, its homogeneity, uniformity and isotropic characteristics are excellent. In addition, the lot-to-lot uniformity has been shown to be very good. Its useful temperature range is greater than any of the other 'pure' materials, and it exhibits very good dimensional stability with temperature. Its humidity resistance is excellent and the cold flow characteristics are negligible.

PPO is attacked by most hydro-carbon chemicals and solvents, but is resistant to most acids, alkalis, and water base solutions. Its mechanical strength and thermal characteristics are good; however, most of the processing restrictions applicable to polystyrene are in effect for PPO. It cannot be sheared or punched, and even improper machining can create problems. PPO is sensitive to an extreme extent to the processing methods used to manufacture the stripline circuit boards. Unfortunately, many of the problems generated by improper handling of the material are not evident until after temperature cycling and/or high power applications and a period of time has passed.

The specific manifestation is stress crazing, that is, small internal cracks which tend to grow throughout the material. They have no electric effect unless they reach the surface and are great enough to rupture one of the copper conductors; in which case, circuit failure occurs. The manufacturers claim, and indeed are correct, that this fault is the result of not heeding precise processing instructions provided with the material. Sufficient to say that PPO has many highly desirable properties, but cannot be abused as many of the other materials can. Techniques such as punching and shearing cannot be used under any conditions. Machine tool speed must be controlled and feed speeds maintained at a rate which does not induce stress. The use of coolants, solvents and degreasers, etc., must be restricted to those specifically recommended. When these precautions are observed, the material is stable, uniform, reliable and highly recommended, particularly for the frequency range below 5.0 GHz.

Data Summary for Polyphenelene Oxide (PPO)

Dielectric constant 2.55
Dissipation factor 0.0016
Useful temperature range –270°F to +375°F
Tensile strength 6000 – 10,500 psi
Flexural strength 15,000 psi
Impact resistance 1.5 ft. lbs./inch
Flexibility – good
Thermal conductivity 4.5 x 10^{-4} cal/sec./cm^2/°C/cm
Coefficient of thermal expansion 2.9 x 10^{-5} inches/inches/°F
Ratio of dielectric to copper thermal expansion 2.1
Thickness range available 0.009" to 0.750"

Ceramic-Filled Resins

Several major manufacturers offer materials which are a mixture of some resin with a filler of ceramic powder. The prime advantage of these products is the wide range of dielectric constants which are available both as standard materials and as specially compounded materials. This flexibility is made possible by controlling the percentage mixture of resin versus ceramic. As a result, the other properties of materials also vary greatly. Polyolefin, polystyrene and various aromatic and aliphatic resins as well as polyester and silicone resin base materials are employed. In general, dielectric constants of 1.7 to 15 are available with most of the standard materials grouping between 2.0 and 6.0. Loss tangents vary from 0.0005 to 0.005, more as a function of the base resin than of the ceramic. However, it should be remembered that a high dielectric constant material will have a greater loss per inch than a low dielectric constant material with the same loss tangent. Thus, most of the higher dielectric constant materials do exhibit substantial losses in practical use.

Production methods have been refined to a point where the uniformity and homogeneity of this class of material is generally excellent. Similarly, the materials exhibit isotropic properties. Dimensional stability varies with the base resin, but, in general, it is better than the stability of the unloaded resin. Temperature ranges are very close to the ranges permissable for the base resins and most of the processing restraints applicable to the resins apply to the mixtures. The data summary presented has a wide range of values and is intended as a general guide only. The user should consult the various manufacturers of these materials in order to determine the exact properties available for any single mixture.

Data Summary for Ceramic Filled Resins

Dielectric constant range 1.7 – 25
Dissipation factor range 0.0005 to 0.005
Useful temperature range –80° to +515°F
Tensile strength 2000 to 7000 psi
Flexural strength 6,000 to 12,000 psi
Impact resistance – poor
Flexibility – poor
Thermal conductivity – approximately 4×10^{-3}
$\qquad\qquad$ calories/second/cm^2/°C/cm
Coefficient of thermal expansion 22 to 32×10^{-6} inches/inch/°F
Ratio of dielectric to copper thermal expansion 1.67 to 2.3
Thickness range available 1/16" to 2"

Quartz-Loaded Teflon

Perhaps one of the best all around materials is woven quartz-reinforced teflon. It is similar to the teflon fiberglass with the exception that its electrical properties are greatly enhanced by the substitution of quartz for glass in the reinforcing cloth. Its dielectric constant is 2.47 and it is well controlled. This, coupled with a loss tangent of 0.0006, make the material far superior electrically to other materials (except polyolefin). Unlike polyolefin, however, the mechanical properties are for the most part far more stable. Its useful temperature range is −80°F to +500°F with dimensional stability as good as or better than that of teflon fiberglass. It is impervious to most chemicals and has excellent physical properties, although, like the teflon fiberglass, carbide tools must be used.

Like other woven laminates, it suffers from some anisotropy and mild water absorption due to wicking. Its bond strength is very high at 7 lbs./inch. In general, quartz teflon is a highly desirable material which provides the best mechanical properties of teflon fiberglass with a close approximation of the electrical performance of polyolefin. However, as with most things, one cannot get something for nothing. Quartz teflon laminates cost approximately six times that of teflon fiberglass and other comparable laminates; thus, for most applications their use is limited.

Data Summary for Teflon Quartz (Woven)

Dielectric constant 2.47
Loss tangent 0.0006
Useful temperature range −80°F to +500°F
Tensile strength 18,000 psi
Flexural strength 15,000 psi
Impact resistance 15 ft./lbs. per inch
Flexibility − good
Thermal conductivity 7×10^{-4} cal/sec./cm^2/°C/cm.
Coefficient of thermal expansion 10.5×10^{-6} inches/inch/°F
Ratio of dielectric to copper thermal expansion .75
Thickness range available 0.010 to 0.187"

Other Materials

There are, of course, other materials which under the correct conditions can be used for stripline applications. For the most part, they offer no significant advantages or are limited for general use by their poor electrical properties. Such a material is G-10 epoxy glass which has a high loss although it possesses excellent mechanical properties. Its use is limited to the low L-band or UHF region. Other materials such as glass-loaded polyolefin are available, but offer very little compared to the more conventional materials. There are other compounds like XXXP phenolic board which are totally unsuitable for microwave work. Of the many materials available, there are only five or six in general use. These materials form the basis for the large majority of stripline circuit work being done at this time.

In later chapters of this book, a number of design curves will be presented. Where possible, these will be normalized for all dielectrics and thickness ranges. There are, however, many cases in which the computations do not readily permit this degree of normalization. For these cases, specific design or performance curves will be presented for the most widely used materials and thicknesses. Specifically, these are:

(1)   Woven teflon fiberglass-
        dielectric constant 2.55

(2)   Microfiber teflon glass-
        dielectric constant 2.40

(3)   Polyolefin-
        dielectric constant 2.32

(4)   Polyphenelene oxide-
        dielectric constant 2.55

(5)   Cross-link polystyrene-
        dielectric constant 2.53

(6)   Glass-filled polystyrene-
        dielectric constant 2.62

Additionally, standard thicknesses of 0.031, 0.062, and 0.125" which provide ground plane spacing of 0.062, 0.125, and 0.250" will be plotted. Where multilayer construction is appropriate, shim material thicknesses of 0.005, 0.011, and 0.022" will be shown with the above base material dimensions. Thus, the reader may make use of the design information provided without recourse to special materials or dimensions.

LINE LOSSES

The attenuation of microwave signals in solid dielectric stripline is a function of a variety of line parameters, including:
    (1)   Characteristic impedance
    (2)   Dielectric constant
    (3)   Loss tangent
    (4)   Conductor resistivity
    (5)   Surface finishes
    (6)   Conductor thicknesses
    (7)   Circuit configurations

In general, however, the losses may be expressed by equations 1–1 and 1–2.[3]

$$a_{TOTAL} = a_{conductor} + a_{dielectric} \qquad (1\text{–}1)$$

$$a_d = \frac{27.3 \sqrt{\epsilon} \quad \text{Tan } \delta}{\lambda} \quad \text{(dB/unit length)} \qquad (1\text{–}2)$$

where

$$\lambda = \text{free space wavelength}$$
$$\epsilon = \text{dielectric constant}$$
$$\text{Tan } \delta = \text{loss tangent of the dielectric}$$

Conductor losses may be determined from the relation

$$a_c = \frac{.0231 \quad R_s \quad \sqrt{\epsilon}}{Z_0} \left[\frac{\delta Z_0}{\delta b} - \frac{\delta Z_0}{\delta w} - \frac{\delta Z_0}{\delta t} \right] \qquad (1\text{–}3)$$

where

Z_0 = characteristic impedance
b = ground plane spacing
w = stripwidth
t = conductor thickness
R_s = conductor resistivity (ohms/square)

This may be simplified for two cases of construction.
(1) The narrow center strip case where the following conditions apply.

a. $\frac{w}{(b-t)}$ is greater than or equal to 0.35

b. $\frac{t}{b}$ is less than or equal to 0.25

c. $\frac{t}{w}$ or $\frac{w}{t}$ is less than or equal to 0.11

(2) The wide center strip case where

a.   little or no fringe field interaction occurs and

b. w/b is greater than or equal to 0.35

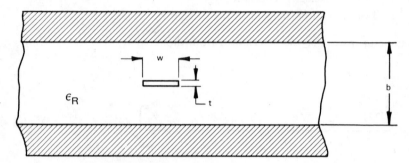

FIG.1–1   Typical Stripline Cross Section

For the narrow strip case, equation 1–4 applies.

$$a_c = \frac{.011402 \sqrt{\epsilon} \sqrt{f_{(GHz)}}}{\sqrt{\epsilon} \ Z_0 \ b} \left[1 + \frac{b}{d_0} \left[.5 + .669 \frac{t}{w} - 0.225 \frac{t}{w} + .5 \pi \log_e \left(\frac{4 \pi w}{t} \right) \right] \right] \quad (1-4)$$

where

d_0 is the equivalent circular cross section[4][5]

For most work, we are more concerned with the wide center strip case. It is this configuration which covers the bulk of the transmission lines used in practice, since for most normal dielectrics, this condition occurs at about $Z_0 = 70$ ohms. In this case where the fringe fields are not a significant factor, the conductor losses may be computed from the relationship,[4]

$$a_c = \frac{2.02\ (10^{-6})\ \sqrt{\epsilon}\ \sqrt{f_{(GHz)}}\ (\sqrt{\epsilon}\ Z_0)}{b} \left[\frac{1}{1-t/b} + \frac{2\ w/b}{(1-t/b)^2} + \frac{1+t/b}{\pi\ (1-t/b)^2} \log_e \left(\frac{\frac{1}{1-t/b}+1}{\frac{1}{1-t/b}-1} \right) \right] \qquad (1-5)$$

Equations 1–1, 1–2, and 1–5 have been tabulated for the standard materials listed in the previously mentioned thicknesses for a 50 ohm line. Calculations were made for both one ounce and two ounce center conductors. The variation was exceptionally small; for example, for 0.125 ground plane spacing polyolefin at 5.0 GHz the total loss for one ounce copper equals 0.0470 dB/inch. For two ounce copper the loss equals 0.0457 dB/inch for a total difference of 0.0013 dB/inch. Inasmuch as most users favor the lighter grade of copper for improved etching and tolerance control, the curves of 50 ohm line losses are presented for the one ounce case. The reduction of about 0.001 to 0.007 dB for 2 oz. copper can be applied with reasonable accuracy over the frequency bands. This data is shown in Figures 1–2 through 1–7.

POWER HANDLING CAPABILITY

Stripline construction is not fundamentally suited to high power applications of either high peak power or CW (average power). This is not to suggest that it is solely limited to milliwatt applications. Properly designed, it is capable of reasonable power levels although it will never approach the capabilities of waveguide or even coaxial lines of comparable cross-sections. This is due to several factors such as the field concentration at the sharp edges of the line and the dielectric losses of the base materials. Peak power problems and average power considerations must be examined as almost unrelated problems.

Peak Power

The failure mechanism of peak power is arcing or "breakdown". The conditions which precipitate this are wide and varied. They include field

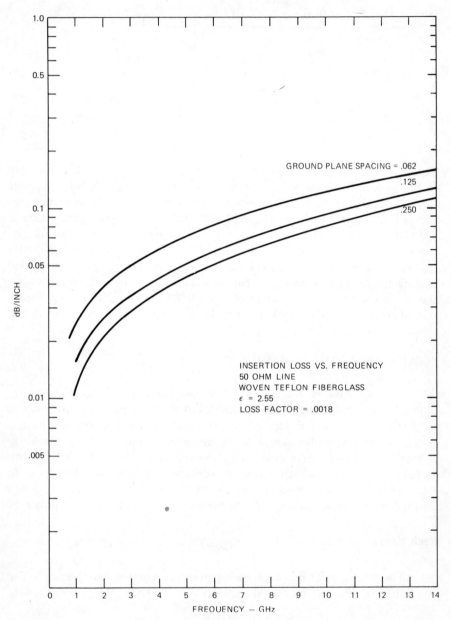

FIG.1–2 Insertion Loss vs. Frequency for Woven
Teflon Fiberglass 50 Ohm Line

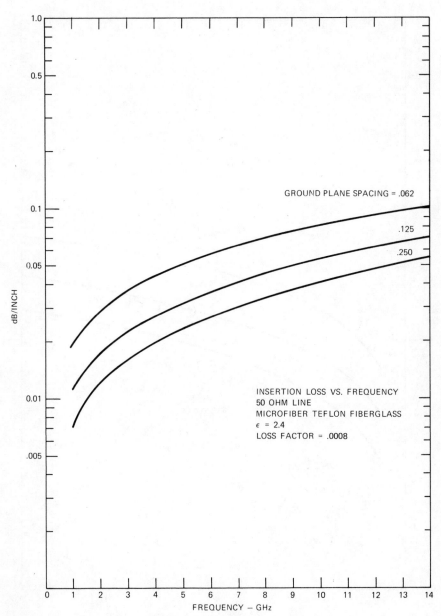

FIG.1–3   Insertion Loss vs. Frequency for Microfiber
Teflon Fiberglass
50 Ohm Line

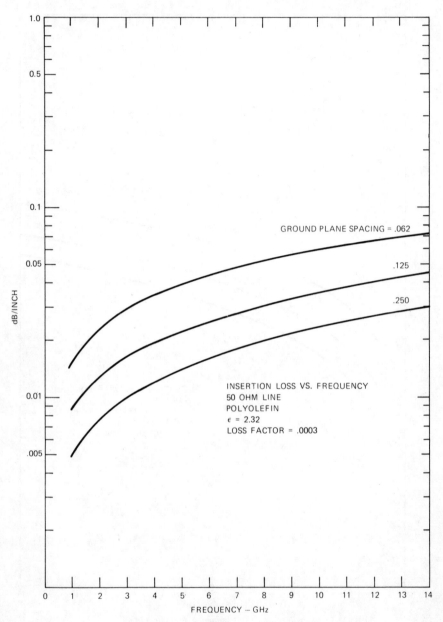

FIG.1–4   Insertion Loss vs. Frequency for
Polyolefin, 50 Ohm Line

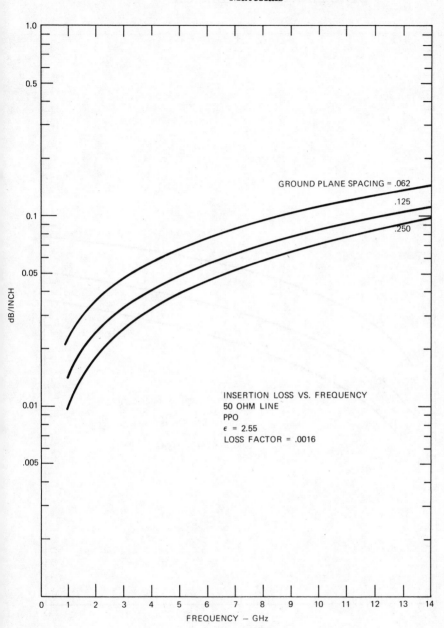

FIG.1–5   Insertion Loss vs. Frequency for Polyphenelene
Oxide, 50 Ohm Line

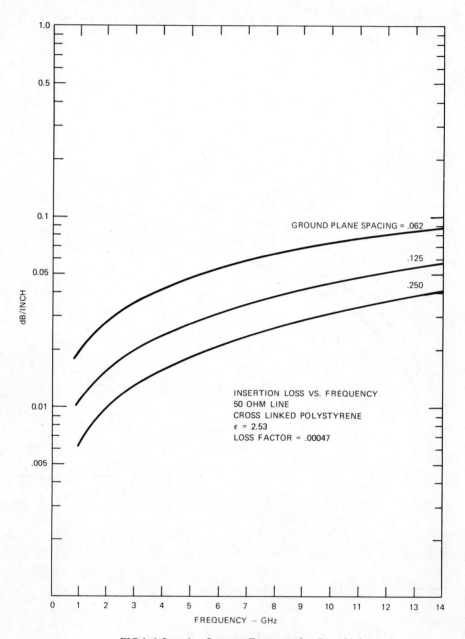

FIG.1–6 Insertion Loss vs. Frequency for Cross-Link
Polystyrene, 50 Ohm Line

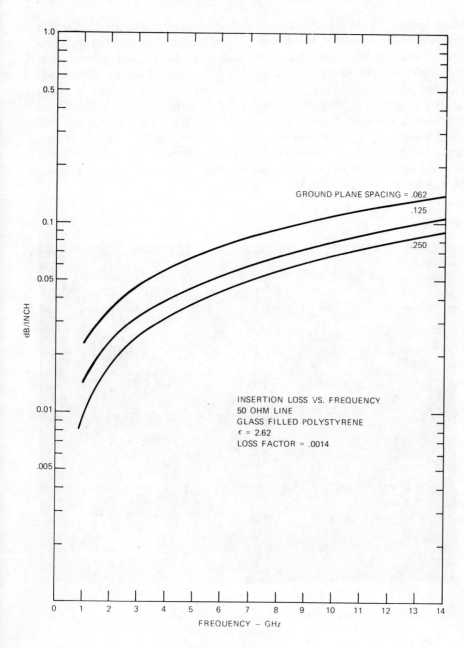

GROUND PLANE SPACING = .062

.125

.250

INSERTION LOSS VS. FREQUENCY
50 OHM LINE
GLASS FILLED POLYSTYRENE
$\epsilon = 2.62$
LOSS FACTOR = .0014

dB/INCH

FREQUENCY – GHz

Filled Polystyrene,
50 Ohm Line

concentrations due to configuration as well as air gaps and sharp edges which serve as field concentrators. The dielectric strength of the base material also plays a part. Data[6] has been presented for the case of rounded center conductors in an air dielectric. This presentation indicates that an air line having a cross section roughly equal to that of 1/8" GPS 50 ohm teflon fiberglass line (but in air) is capable of handling about 50 kilowatts for a perfect match. It should be remembered, however, that this is calculated for a homogeneous medium with rounded edges on the center conductor. In practice, the situation is far less than ideal. The edges of photo-etched stripline are fairly sharp, although some rounding occurs due to the undercut and the resist breakdown on the line during the etching process. This can be seen in Figure 1–8.

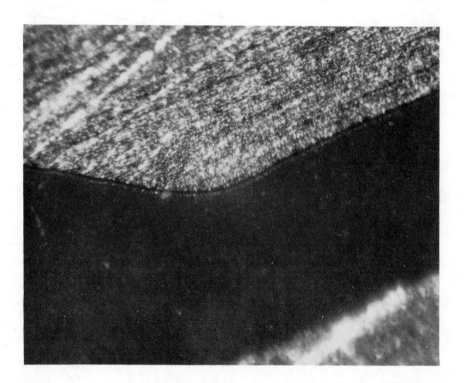

FIG.1–8 Photograph of Etched Line Illustrating
Corner Roundness and Undercut

An additional problem occurs because of the air gap at the edge of the line due to the line thickness and the inability of most base materials to completely deform and flow over the edges. This has been completely eliminated by several manufacturers who now offer laminating services. This technique makes use of a thin sheet of some soft dielectric material such as polyethylene which is placed between the etched board and the cover board and then laminated in place with heat and pressure. Such a laminate is far more resistive to high power breakdown than a normally constructed line or even one in which substrate fillers such as silicone grease have been added to close up the air gaps.

Peak power data is scarce and difficult to calculate accurately. One report,[7] has indicated that 25 kW for a 50 ohm line in 1/8" ground plane spacing using a dielectric of cross-link polystyrene is a safe power level. The author has observed measured performance of 25 kW in 1/8" ground plane spacing using teflon fiberglass in C-band. 40 kW at L-band using 1/4" ground plane spacing has also been observed. Beyond this, very little is known. One point of practical interest, however, is that in most cases the 3 mm connectors and their transitions in common use today will break down before the line itself, except in extreme cases such as tightly coupled lines or extreme junction discontinuities. In general, 5 kW seems a safe level for well-constructed line. Improvements may be made by various laminating techniques and careful design and fabrication, however, the limits are very much dependent upon fabrication practice and cannot be reliably predicted. One additional factor, which may significantly reduce the power handling capability, is the effect of internal mismatches, inasmuch as high VSWR's can produce points of peak power far in excess of the line power within a package resulting in breakdown of these points at levels which might not otherwise be expected for a well-constructed line.

Average Power

The average power capability of stripline is primarily a function of the permissible temperature rise of the center conductor and surrounding laminate. It is, therefore, related to the dielectric used, its thermal conductivity and electrical loss, the cross section of the line, any case or supporting material, the maximum allowable temperature and the ambient temperature. Thermal energy in a stripline application behaves in a fashion very similar to that of the electrical and magnetic field distribution of the transmitted RF energy.[8][9] This may be seen in Figure 1–9.

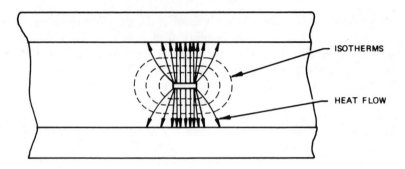

FIG.1–9   Thermal Characteristics of a Stripline Showing
the Similarity to the Electric and
Magnetic Field Characteristics of the
Same Type Line

It has been suggested[9] that the dielectric losses are actually generated midway between the center conductor and the ground plane, thus reducing the impact on center conductor heating. It would appear however, that these losses probably are concentrated nearer the center conductor as a result of the field concentration, and that as such they are effectively at the center line in terms of their thermal concentration. Since radiation occurs towards both ground planes and while the path may be shorter toward one side, it is of necessity, then, longer toward the other. A simple technique[10] which makes use of the concept of thermal resistance permits a reasonable calculation of the temperature rise of the stripline center conductor. If we take each of the interfaces of conventional line and determine the temperature drop from interface to interface as shown in Figure 1–10, then the thermal rise of the center conductor can be calculated from equation 1–6.

$$\Delta T_{TOT.} = \Delta T_{(1\text{-}2)} + \Delta T_{(2\text{-}3)} + \Delta T_{(3\text{-}4)} \qquad (1\text{-}6)$$

where

$$\Delta T_{(N,\ N+1)} = (1/K)\,(L/A)\,Q \qquad (1\text{-}7)$$

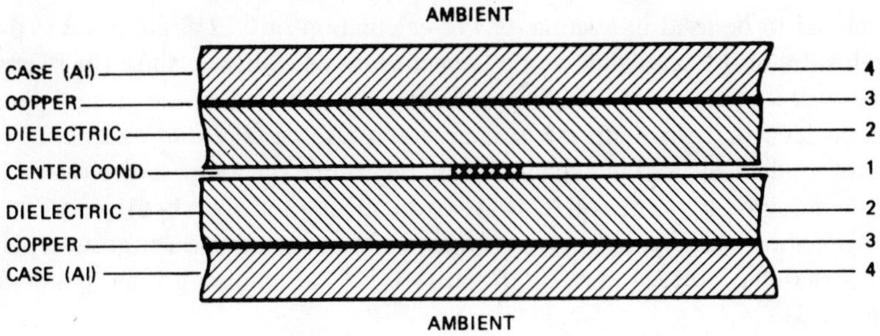

AMBIENT

CASE (Al) ——— 4
COPPER ——— 3
DIELECTRIC ——— 2
CENTER COND ——— 1
DIELECTRIC ——— 2
COPPER ——— 3
CASE (Al) ——— 4

AMBIENT

FIG.1–10   Cross-Section of Line Construction Used
for the Thermal Analysis Curves

and

K	=	thermal conductivity
A	=	cross section of heat flow path
L	=	path length
Q	=	heat flow

If we let $1/K$ = thermal resistivity = R in units of °C inches/watts then equation 1–7 can be expressed as

$$\triangle T_{(N,\ N\ +1)} = \frac{(R)\ (L)\ (P)}{A} \qquad\qquad (1\text{–}8)$$

where

P   =   watts dissipated.

   Since very few components are built without some form of case or mechanical support in addition to the actual board material, it was decided that for a model to be used in average power calculations a 0.125" aluminum cover on both sides of the sandwich would be incorporated. Further, since the thermal field, as well as the electric field, is broadening as it approaches the ground planes it was decided that this increase in area would also be taken into account. The resultant model was used in calculations to determine the temperature rise of the center conductor in terms of °C/watt of incident (not absorbed) power. The results are shown in Figures 1–11, 1–12, and 1–13 which represent ground plane spacings of 0.062", 0.125" and 0.250" with a 0.125" aluminum ground plane case on each side.

Several interesting observations may be made from these curves. The first is that there is no one material which is superior under all conditions in terms of power capabilities. It can be seen that since the copper loss has contributed a substantial portion of the power losses and thus the heating effect the thermal rise does not remain constant regardless of ground plane spacing as might be expected. True, as the ground plane spacing increases, the line area also increases proportionately, thus nullifying the effect on the thermal equations. However, a change in ground plane spacing also has an impact on the actual dissipated power, thus, as a general statement it can be said that a larger ground plane

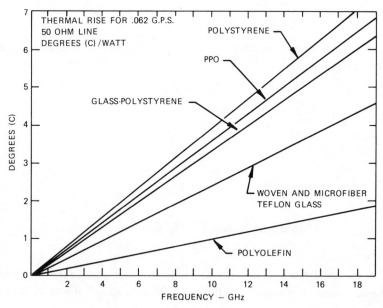

FIG.1–11   Temperature Rise .062 to Ground Plane
Spacing, 50 Ohm Line

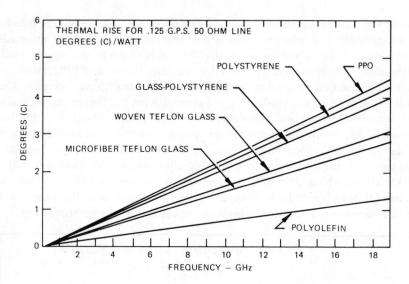

FIG.1-12 Temperature Rise for 0.125 Ground Plane
Spacing, 50 Ohm Line

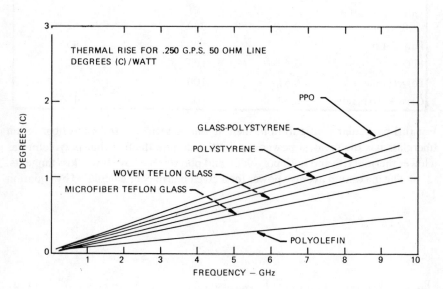

FIG.1-13 Temperature Rise for 0.250 Ground Plane
Spacing, 50 Ohm Line

spacing will handle a greater amount of average power. This is always true for a given material, but is not necessarily true throughout the range of materials. Examination of the curves will show that for a ground plane spacing of 0.062", the thermal rise of cross-linked polystyrene is greater than that of PPO. However, when the ground plane spacing increases to 0.125" the situation reverses. This is due to the fact that the smaller ground plane spacing has higher copper losses, thus increasing the power dissipated in a material with poorer thermal characteristics. When the ground plane spacing increases, the dominant losses are due to the dielectric; thus the polystyrene with its lower loss proves better than the PPO despite the superior thermal characteristics of the latter. Other considerations which determine the maximum average power level are the ambient temperature and the maximum dielectric temperature, thus, the lowest thermal rise material may not necessarily be capable of the highest power.

MATERIAL	ΔT/W	MAXIMUM AVERAGE POWER .125 GPS 50 OHM LINE 25°C at 2.5 GHz	
		MAXIMUM OPERATING TEMPERATURE °C	MAXIMUM AVERAGE POWER (watts)
Woven teflon fiberglass	0.42	265	560
Microfiber teflon fiberglass	0.38	260	620
Polyolefin	0.22	100	455
PPO	0.60	194	324
Polystyrene	0.57	100	175
Glass Polystyrene	0.54	100	185

For this particular configuration, then, it can be seen that the microfiber teflon fiberglass has the greatest power capabilities despite the fact that polyolefin has a lower thermal rise and both polyolefin and polystyrene have lower loss tangents.

The maximum average power for a given line may be calculated by equation 1-9.

$$P_{MAX.} = (T_{MAX.} - T_{AMB.})/\Delta T \qquad (1-9)$$

Comparison of maximum power calculated from the curves with previously published data[11] will show these figures to be somewhat more conservative. This is probably due to the use of the outer plates in the calculation as well as slightly more conservative estimates of loss tangents.

Conclusion

The properties of the laminates chosen for a given design problem will have a significant effect on the success or failure of the end product. Mechanical characteristics, so often ignored, are a major portion of any electrical design, since its realizability and stability in a working environment depends upon them. Obvious solutions are not always the correct ones since all the factors, such as dielectric constant, loss tangent, thermal properties, dimensional stability, homogeneity, etc., must be evaluated for each situation. The result of this evaluation may be a determination that stripline may not be the correct medium for the job. If it is, however, the design data presented here should be of value.

REFERENCES

1   Vossberg, W. A., *Stripping the Mystery from Stripline Laminates,* Microwaves, Vol. 7, No. 7, January 1968, pp. 104–110.

2   Howe, H., *Dielectrically Loaded Stripline at 18.0 GHz,* Microwave Journal, Vol. 9, No. 11, January 1966, pp. 52–54.

3   Cohn, S., *Problems in Strip Transmission Lines,* MTT-3; No. 2, March 1955, pp. 119–126.

4   Ibid.

5   *Microwave Engineers Handbook and Buyers Guide,* Horizon House, 1966, p. 126.

6   Ibid, p. 128.

7   Peters, R. W., et al, *Handbook of Tri-Plate Microwave Components,* Sanders Associates, Nashua, New Hampshire, 1956.

8   Levine, R. C., *Determination of Thermal Conductance of Dielectric Filled Strip Transmission Line from Characteristic Impedance,* MTT-15, No. 11., November 1967, pp. 645–646.

9   Shiffres, P., *How Much CW Power Can Stripline Handle,* Microwaves, June 1966, pp. 25–34.

10   Kline, G., *Thermal Resistivity Table Simplifies Temperature Calculations,* Microwaves, February 1970, pp. 58–59.

11   *Microwave Engineers Handbook and Buyers Guide,* Horizon House, 1966, p. 128.

CHAPTER

—2

Characteristic Impedance, Launching Methods, and Basic Structures

Accurate determination of the characteristic impedance of the transmission line used in component design is essential. The bibliography attests to the interest and number of methods used for this calculation by the large number of sources available on this subject. Exact solutions for all possible cases are extremely cumbersome and the value of the small increase in accuracy over reasonable and valid approximations is questionable. The most widely used method of calculation is a set of equations reported by Cohn.[1]

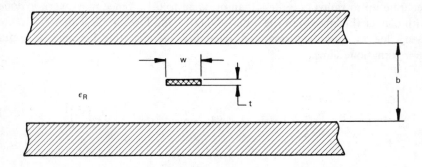

FIG. 2-1 Stripline Cross-Section Establishing Electrical and Mechanical Dimensions

These relate to the configuration shown in Figure 2-1. For stripwidth which are very small, i.e., $\frac{w}{(b-t)}$ less than or equal to 0.35 and $\frac{t}{b}$ less than or equal to 0.25, the characteristic impedance is very similar to that of a round wire between flat ground planes as shown in Figure 2-2. The relation for this case is given[2] as:

$$Z_0 = \frac{138}{\sqrt{\epsilon}} \, \log_{10} \, \frac{4\,b}{\pi\,D} \tag{2-1}$$

By establishing an equivalence between a round center conductor and the rectangular one, this equation can be applied.

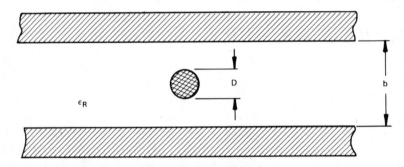

FIG. 2-2   Cross-section of A Round Wire Between
Symmetric Ground Planes

Data has been presented in a graphical form[1] and also as an approximate formula accurate up to ratios of $\frac{t}{w}$ less than or equal to 0.11. Since most work is done with copper thicknesses in the order of 0.0014 and few lines less than 0.012 are used, due to manufacturing and reliability reasons, this becomes a useful approximation. Thus,

$$d = \frac{w}{2} \left[1 + \frac{t}{w} \left(1 + \log_e \frac{4\,\pi\,w}{t} + .51\,\pi \left(\frac{t}{w} \right)^2 \right) \right] \tag{2-2}$$

$$Z_0 \, \sqrt{\epsilon} = 60 \, \log_e \left(\frac{4\,b}{\pi\,d} \right) \tag{2-3}$$

If the strip width is wide enough so that the fringe fields do not interact, i.e.; w/b greater than or equal to 0.35, then another relation is applicable

$$Z_0 \sqrt{\epsilon} = \frac{94.15}{\left(\dfrac{w/b}{1 - t/b} + \dfrac{C'_f}{0.0885 \, \epsilon_R} \right)} \tag{2-4}$$

where,

$$C'_f = \frac{0.0885 \, \epsilon}{\pi} \left[\frac{2}{1 - t/b} \, \log_e \left(\frac{1}{(1 - t/b)} + 1 \right) - \left(\frac{1}{1 - t/b} - 1 \right) \log_e \left(\frac{1}{(1 - t/b)^2} - 1 \right) \right] \text{Pf/cm} \tag{2-5}$$

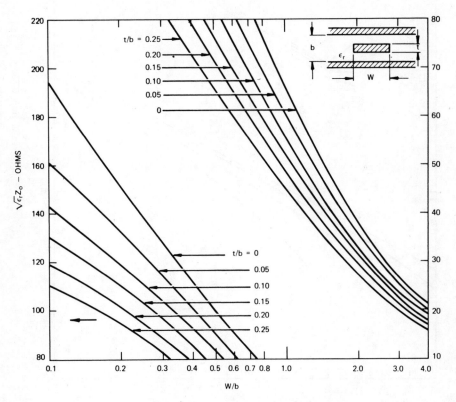

FIG. 2–3   General Curves for Characteristic Impedance
of Dielectrically Loaded Stripline

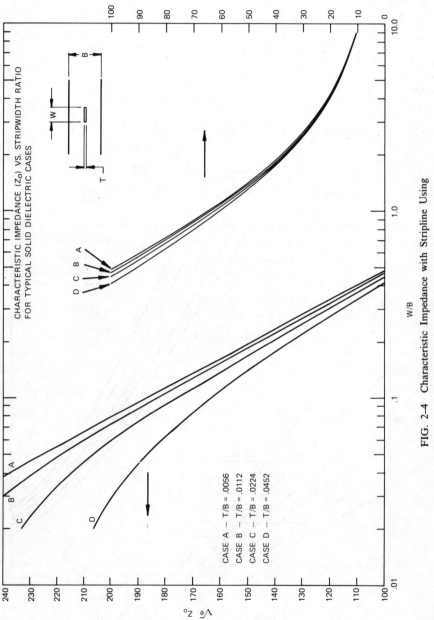

FIG. 2–4 Characteristic Impedance with Stripline Using Commonly Available T/B Ratios

These two equations for Z_0 are accurate to better than 1.3%, the greatest error occurring at the point where $\frac{w}{(b-t)}$ equals 0.35. Their values have been plotted for a number of cases and are presented in Figure 2–3[3].

Since the cases of $\frac{t}{b}$ most commonly encountered in printed solid dielectric stripline are not included in this general figure, they have been plotted separately in Figure 2–4.

Case A represents one ounce copper with a 0.250" ground plane spacing. Case B represents one ounce copper with a 0.125" ground plane spacing, or two ounce copper with a 0.250" ground plane spacing. Case C represents one ounce copper with a 0.062" ground plane spacing, or two ounce copper with a 0.125" ground plane spacing and Case D represents a two ounce copper with a 0.062" ground plane spacing. These curves are normalized and may be used for materials with any dielectric constant. Their accuracy is approximately 1.5% and, in most cases, is better than the fabrication tolerances achievable in normal production.

For many of the constructional techniques presented in the later chapters, it will be desirable to use a three-layer configuration. This will cause the center conductor to be offset by half the thickness of the central layer as shown in Figure 2–5.

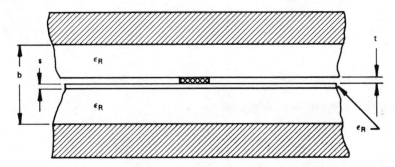

FIG. 2–5   Cross-Section of an Offset Center Conductor
Strip Transmission Line

This will cause an unbalance in both parallel plate capacities as well as the fringe field capacities which can result in an error of several ohms in the calculation of the 50 ohm line. A method of calculation which works well was developed by Rosenzweig[4], and is presented here. Use is made of the relation of characteristic impedance to shunt capacity of the line; thus,

$$Z_0 \sqrt{\epsilon} = n/(c/\epsilon) \text{ ohms} \tag{2-6}$$

where   n = free space impedance = 376.7 ohms

$\dfrac{c}{\epsilon}$ = is the ratio of the static capacitance per unit length between conductors to the permittivity of the dielectric medium

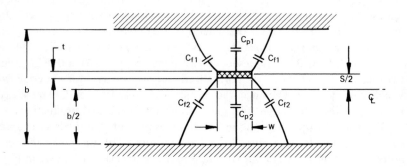

FIG. 2-6   Design, Dimensions,and Characteristics for
Offset Center Conductor
Strip Transmission Line

The total capacity then becomes:

$$(c/\epsilon)_{TOT} = \left[C_{P1}/\epsilon + C_{P2}/\epsilon + 2\,C_{f1}/\epsilon + 2\,C_{f2}/\epsilon \right] \tag{2-7}$$

for the case of $\dfrac{w}{(b-t)}$ is greater than or equal to 0.35

$$C_p/\epsilon = \frac{2\,w/b}{1 - t/b} \tag{2-8}$$

therefore,

$$C_{P1}/\epsilon = \frac{2\,w/(b-s)}{1 - t/(b-s)} \tag{2-9}$$

and

$$C_{P2}/\epsilon = \frac{2\,w/(b+s)}{1 - t/(b+s)} \tag{2-10}$$

Similarly,

$$C_{f1}/\epsilon = \frac{1}{\pi} \left[\frac{2}{1 - t/(b-s)} \, \log_e \left(\frac{1}{1 - t/(b-s)} + 1 \right) - \left(\frac{1}{1 - t/(b-s)} - 1 \right) \, \log_e \left(\frac{1}{(1 - t\,(b-s))^2} - 1 \right) \right] \tag{2-11}$$

$$C_{f2}/\epsilon = \frac{1}{\pi} \left[\frac{2}{1 - t/(b+s)} \, \log_e \left(\frac{1}{1 - t/(b+s)} + 1 \right) - \left(\frac{1}{1 - t/(b+s)} - 1 \right) \, \log_e \left(\frac{1}{(1 - t/(b+s))^2} - 1 \right) \right] \tag{2-12}$$

Substitution of these values into equation 2–6 will provide accurate results as long as $\frac{w}{(b-t)}$ is greater than or equal to 0.35. A correction for the case where $\frac{w}{(b-t)}$ is less than 0.35 may be made by the method of Getsinger.[5] Thus, we can define a new value of w/b:

$$\frac{W_N}{b} = \frac{(0.07\,(1 - (t/b)) + w/b)}{1.2} \tag{2-13}$$

which provides reasonable correction to the fringing capacity for the condition

$$0.1 < \frac{W_N/b}{(1 - t/b)} < 0.35 \tag{2-14}$$

This technique has been used to generate the curves of Figures 2–7, 2–8, and 2–9 which represent base materials 0.031", 0.062", and 0.125" thick with shim thicknesses of 0.011", 0.022", 0.031", and 0.062". As in the case of the symmetrically located center conductor, these curves are normalized and may be used with any dielectric material.

Connectors and Launchers

While stripline is a useful medium for component design, it is not a normal transmission line medium for use in power transmission or system interconnection. Therefore, in almost all cases, it is necessary to transform from either waveguide or coaxial line into the stripline configuration. Waveguide transitions are used less frequently than coaxial ones since most stripline is used either for low power receiving applications or broadband systems where the bandwidth of conventional waveguides is not adequate. For those cases where waveguide transitions are used, the normal procedure is to simulate the waveguide to coaxial line transducer which consists of a probe, either coaxial or, in some cases, printed, which extends through the broad wall of the waveguide with a tuning short in the guide approximately a quarter-wave behind it. The actual design dimensions are usually determined empirically. Three such transitions are shown schematically in Figure 2–10.

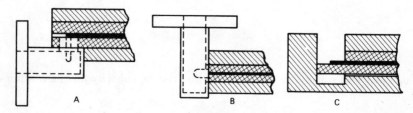

FIG. 2–10   Typical Methods of Launching from
Waveguide to Stripline

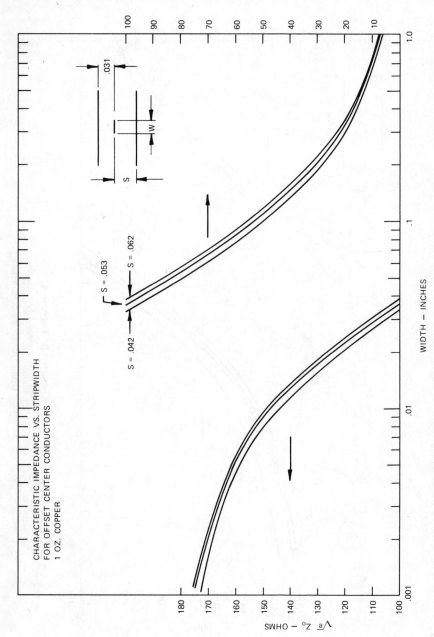

FIG. 2-7 Characteristic Impedance vs Stripwidth for Offset Center Conductors Using 0.031" Base Material

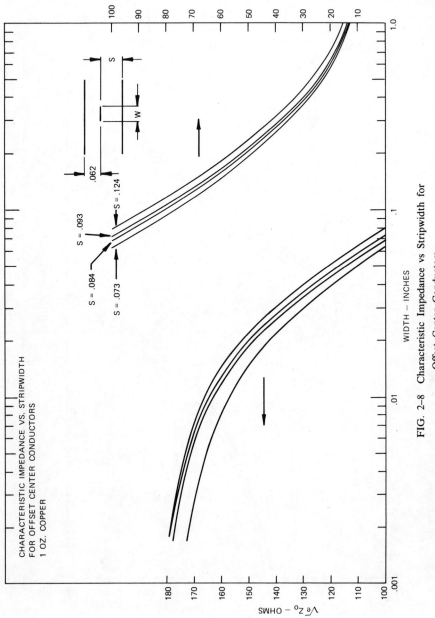

FIG. 2-8 Characteristic Impedance vs Stripwidth for
Offset Center Conductors
Using 0.062" Base Material

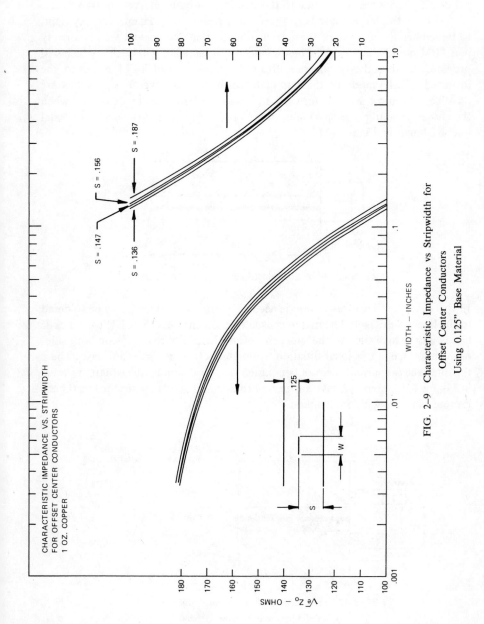

FIG. 2–9 Characteristic Impedance vs Stripwidth for
Offset Center Conductors
Using 0.125" Base Material

Figure 2–10a and 2–10b are coaxial probes which feed the stripline either at right angles or in-line. Figure 2–10c is a completely printed case in which the line is fed directly from the waveguide to the stripline without intervening transitions.

By far the largest number of feeds to stripline are coaxial. The key point to remember in the case of these transitions is that the coaxial feed is already in a TEM mode; therefore, the best connector or launcher will be that which will provide the smoothest transition from the pure coaxial TEM mode to the distorted TEM mode of the flat plate construction. Several approaches are possible. The most straight forward and broadest band technique occurs when the chosen ground plane spacing is close to the outside dimensions of the coaxial line as shown in Figure 2–11.

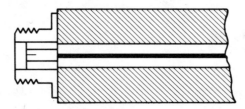

FIG. 2–11   Ideal Coaxial/Stripline Launch Configuration

In this case, no transition section is needed. A simple tab which can be soldered to the line is sufficient. The match is usually excellent (less than 1.1) over a large bandwidth. This is due to the similarity of mechanical dimensions on both sides of the interface in this ideal situation. Unfortunately, there are many cases where the interface dimensions are not compatible, as, for example, the situation shown in Figure 2–12. Here we have an attempt to mate a Type N connector to a 0.125" ground plane spacing in stripline.

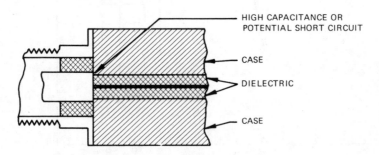

FIG. 2–12   Interface Cross-Section for Large Diameter
Coaxial Lines and Narrow Ground
Plane Spacing Striplines

The problem occurs as a result of the fact that the type N center conductor is 0.120" in diameter, thus creating a very high fringing capacity to the ground planes at the interface, and in fact, a possible short circuit should there be the slightest misalignment. The solution here is to provide an intermediate section of either coaxial line, stripline, slab line or a mixture. These alternates are shown in Figure 2–13.

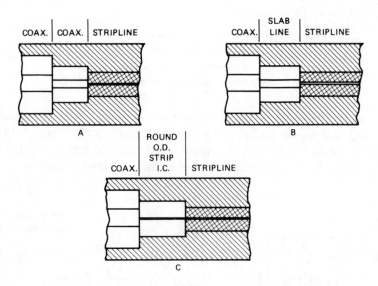

FIG. 2–13   Several Transitions from Large Coaxial to Small
Coaxial to Small Ground Plane
Spacing Stripline

As in the case of coaxial line steps, some adjustment will be necessary to compensate for the fringe capacity of the step faces. One technique, which is very effective, is to adjust the diameters of the inner and outer conductors of the launcher as well as the ground plane spacing so that there are no center conductor steps. It is then possible to launch into the line with a minimum number of interfaces which result in a low VSWR, broadband match. This is illustrated in Figure 2–14.

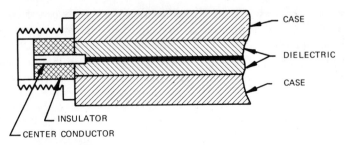

FIG. 2–14   Cross-Section of a Stripline Whose Ground Plane
Spacing Has Been Adjusted To Be
Compatible with the Center
Conductor of the Launching Coaxial Line

There are also situations in which it is convenient to launch coaxially into the stripline from the broadwall rather than from the edge. Figure 2–15 shows such a configuration. Figure 2–15a illustrates the method of mounting and Figure 2–15b shows the typical hole pattern of mounting screws or eyelets.

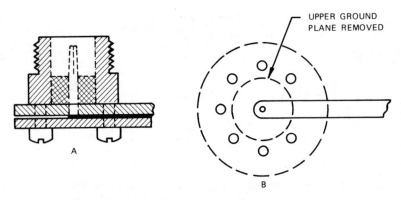

FIG. 2–15   A Simple Technique for Broadwall Launching
a Coaxial Connector to Stripline

These screws serve not only as holding devices, but also as mode suppressors since there is an abrupt change in both direction and mode shape. Broadwall transitions work well in the lower frequency ranges (UHF, L, and S-band). However, they become less perfect at C-band, requiring careful adjustment. At X-band and higher they are virtually useless except for extremely narrow band operation. Even in that case, it is frequently necessary to make Smith chart plots and hand adjust every one because of the large impact of small manufacturing variations such as solder joints. In general, the broadside approach is not recommended at frequencies above 4.0 GHz unless the user is willing to tolerate high VSWR (greater than 1.5) and/or tedious adjustment.

FIG. 2–16   Typical Stripline Coaxial Launchers

Bends and Corners

One of the key advantages of stripline construction is its two dimensional nature which permits the interconnection of many components without the need to break the outer conductor shielding. This same advantage permits the user to place the inputs and outputs with a high degree of flexibility. To do this, a means of changing line direction must be employed. An abrupt right angle bend is mismatched at all frequencies with the magnitude of the mismatch increasing with frequency; thus either a small bend or a compensated corner must be employed.

FIG. 2–17 Photograph of Test Fixture Used to Generate
the Time Domain Study on Right-Angle Bends and Miters

Figure 2–18 shows time domain reflectometer plots for several
configurations. Figure 2–18a is a straight section showing the effect of the
connectors and providing calibration of the 50 ohm line.

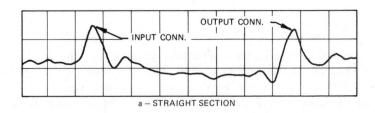

FIG. 2–18 Time Domain Reflectometer Plots of Several
Right-Angle Bends and Miters

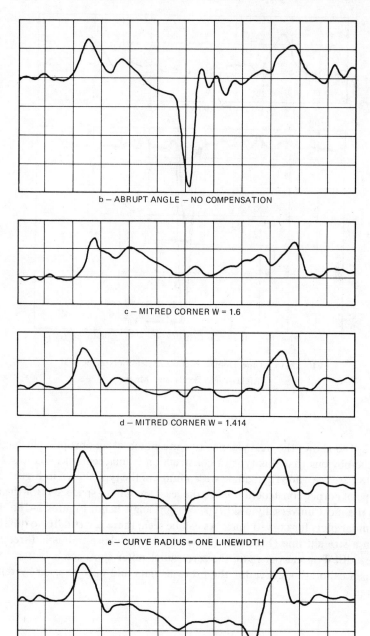

b – ABRUPT ANGLE – NO COMPENSATION

c – MITRED CORNER W = 1.6

d – MITRED CORNER W = 1.414

e – CURVE RADIUS = ONE LINEWIDTH

f – CURVE RADIUS = TWO LINEWIDTHS

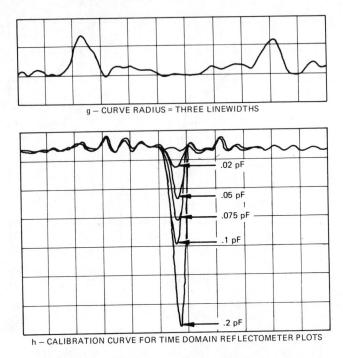

g — CURVE RADIUS = THREE LINEWIDTHS

.02 pF

.05 pF

.075 pF

.1 pF

.2 pF

h — CALIBRATION CURVE FOR TIME DOMAIN REFLECTOMETER PLOTS

FIG. 2–18 Time Domain Reflectometer Plots of Several
Right-Angle Bends and Miters

Figure 2–18b shows the heavy capacitive mismatch of an abrupt right angle. It should be obvious that this type of construction is unacceptable. Figures 2–18c and 2–18d are the same right angle with a miter of the type shown in Figure 2–20. This type of corner construction takes the least amount of space and for the most part, is the most universally used type of right angle bend. Figures 2–18d, f and g are similar plots for curved lines. As can be seen, there is virtually no difference between a straight line (Figure 2–18a) and a curve whose radius is three times the linewidth. For sharper radii, the zero radius miter approach is better. Figure 2–18h is a calibration curve for the previous data and may be used by relating it to Figure 1–19[6].

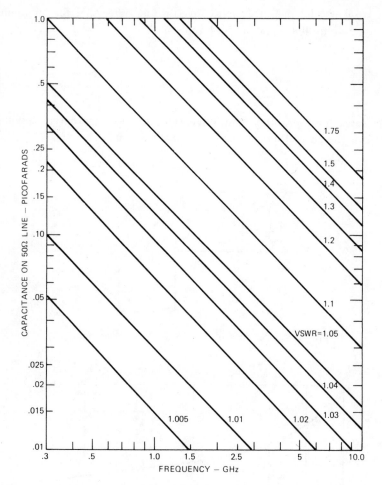

FIG. 2–19   VSWR Generated by a Shunt C on a 50 Ohm
Line as Plotted vs Frequency

It will be noted that the magnitude of the discontinuity for an abrupt angle is not as great as that previously reported.[7] [8] This is due to limitations in the instrumentation used, so Figure 2–18 should be considered more as a qualitative presentation than as a purely quantitative one.

One point which the TDR data does establish is the capacitive nature of the mismatch. Thus the reduction of the surface area of the corner by means of mitering or notching can be employed to reduce the mismatch. Several writers have presented curves and data on the correct dimensions for a miter.[8] [9] [10] For the thin strip in solid dielectric, most investigators agree that for minimum VSWR in a 50 ohm line equation 2–14 applies.

$$W/W_{z_0} = 1.6$$

For other cases, however, other miter dimensions are more appropriate. Figure 2–21 presents the general dimensions for a matched stripline corner.

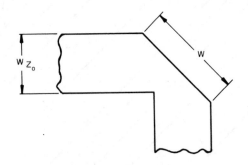

FIG. 2–20   Definition of Right-Angle Miter Dimensions

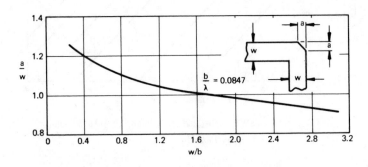

FIG. 2–21   Optimum Designed Dimensions for
Right-Angle Miters

Although stated for a specific $\frac{b}{\lambda}$, the author has found this construction to be fairly insensitive to frequency, requiring additional trimming only at frequencies above C-band.

Shunt Stubs as DC Returns

There are many circuit applications where it is desirable to have the center conductor shorted to the ground plane for dc or very low frequencies without degrading the performance at the design frequency. The simplest technique for achieving this is a single quarter-wave long, high impedance line, short-circuited at the end. The in-band VSWR of this structure is a function of its impedance ratio to the main line and the bandwidth required. As Figure 2–22 shows, the bandwidth for a given VSWR increases as the shunt stub impedance increases.

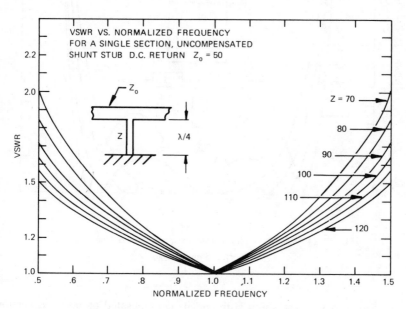

FIG. 2–22 VSWR vs Normalized Frequency for a Single Section Uncompensated
Short Circuited Shunt Stub

Because the dimensional limitation of etched linewidths on plastic-based materials is in the order of 0.010" for reliable design, 100–120 ohms is a practical limit for shunt stub impedance. This results in a limitation of useful bandwidth for this design.

A compensated stub, which makes use of additional quarter-wave, low impedance transformers on each side of the shunt stub is shown in Figure 2–23[11]. The frequency vs VSWR response of this structure is far superior to that of the uncompensated stub. It can be used to improve the match even at narrow bandwidths, provided that there is sufficient room in the circuit to accomodate the additional line lengths of the transformers. Figure 2–24 shows the effect on VSWR, ripple, and bandwidth of varying impedance transformers for a fixed 100 ohm shunt stub.

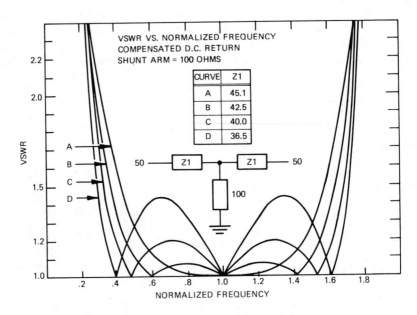

FIG. 2–24   VSWR vs Normalized Frequency for a Compensated Shorted Shunt Stub

For many years, the shunt stub and its compensated broad-band version were the only successful methods used to establish a dc return at high microwave frequencies. Recently, however, new technology has permitted the use of very, very small inductors which can be applied as lumped element devices. These components have the desirable benefit that up to the point of self-resonance they are usable over an extremely broad band of frequencies, thus, their use is potentially possible up through the 18.0 GHz region. Figure 2–25 shows the low frequency response and Figure 2–26 the high frequency response for a number of inductors.

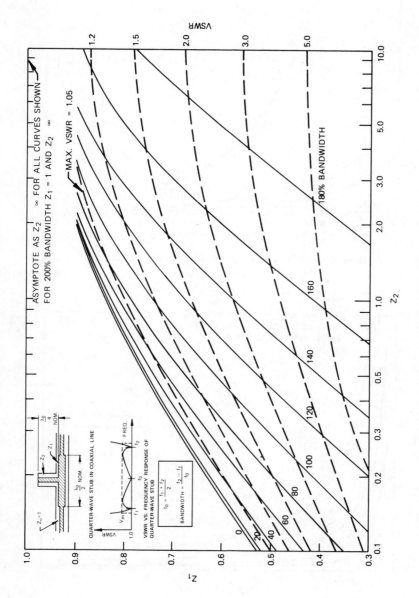

FIG. 2–23   Compensated Shorted Stub Design Curves

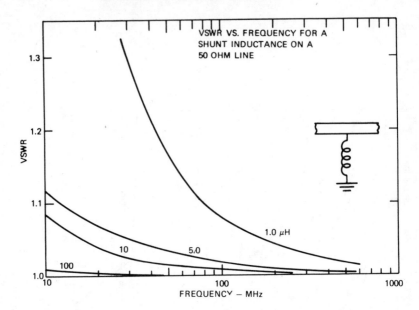

FIG. 2–25   VSWR vs Frequency for a Shunt Inductance
on a 50 Ohm Line

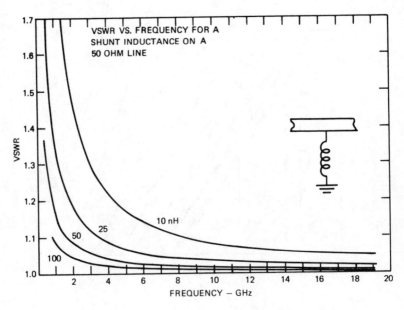

FIG. 2–26   VSWR vs Frequency for a Shunt Inductance
on a 50 Ohm Line

D. C. Blocks

For many of the same reasons that make it desirable to bring the dc potential of the center conductor to ground, it is also frequently necessary to provide a dc and low frequency block in the center conductor either for bias or for I.F. frequency isolation. An effective dc block may be built using a quarter-wave series stub. In a coaxial line, the structure would be as shown in Figure 2–27.

FIG. 2–27   Coaxial Line Series Stub Cross-Section

Translation of this to flat stripline may be accomplished in one of several ways. The most technically correct method is to make use of multi-layer construction as shown in Figure 2–28a.

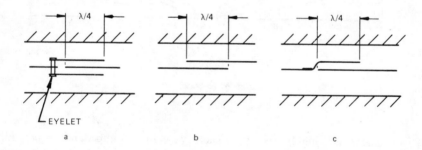

FIG. 2–28   Four-Layer Stripline Series Stub Construction

For this case, a four-layer sandwich is needed. The feed line is split into two lines which are maintained at the same potential by means of interconnecting eyelets. The series stub is completely shielded and its impedance may be readily calculated by means of standard equations and curves such as those in Figures 2–3 and 2–4. Ground plane spacings for this calculation should be the spacing between the two equal potential center conductor elements (b'). This structure is completely equivalent to the coaxial structure provided that the series stub finger is sufficiently shielded from the outside ground plane; thus. it is necessary that its width not be quite as great as the full width of the two sections of the line which are serving as its ground plane. If this condition is met, the structure may be used as a pure series stub, both in filters and in dc block applications.

Frequency response for a single section quarter-wave series stub with impedances varying from 10 to 50 ohms is shown in Figure 2–29.

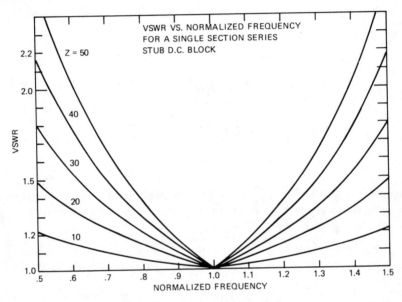

FIG. 2–29 Frequency Response for a Single Section

Quarter-Wave Series Stub

As can be seen, it is highly desirable to achieve as low an impedance as possible in order to broadband the structure. However, in a flat construction line, this is not always possible due to the limitations of the line width of the basic 50 ohm line itself. This limitation can be minimized by using as thin a center-section material as possible so that the ground plane spacing of the internal section of the series stub is very, very small, thus permitting low impedances for relatively narrow stripwidths.

A pseudo-series stub can be built by either the method shown in Figure 2–28b or 2–28c. In the one case, three-layer construction is used while in the other an overlap piece is added on top of a very thin dielectric insulator. By considering

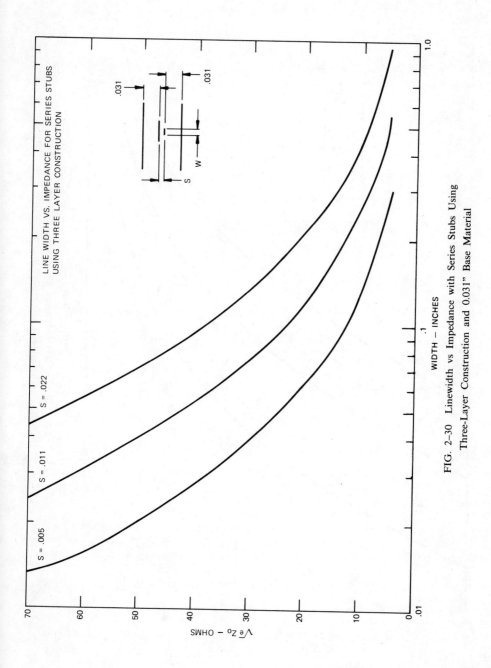

FIG. 2-30 Linewidth vs Impedance with Series Stubs Using
Three-Layer Construction and 0.031" Base Material

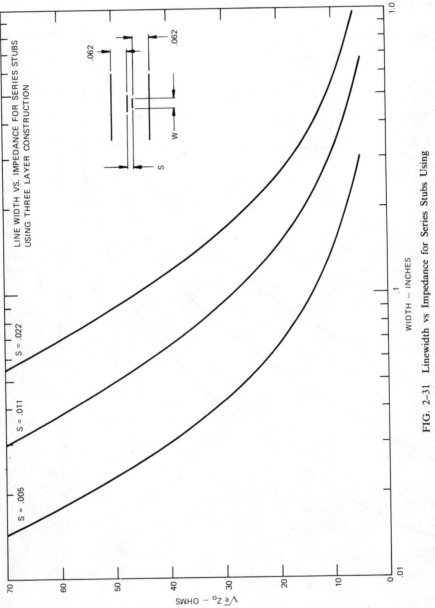

FIG. 2-31 Linewidth vs Impedance for Series Stubs Using Three-Layer Construction and 0.062" Base Material

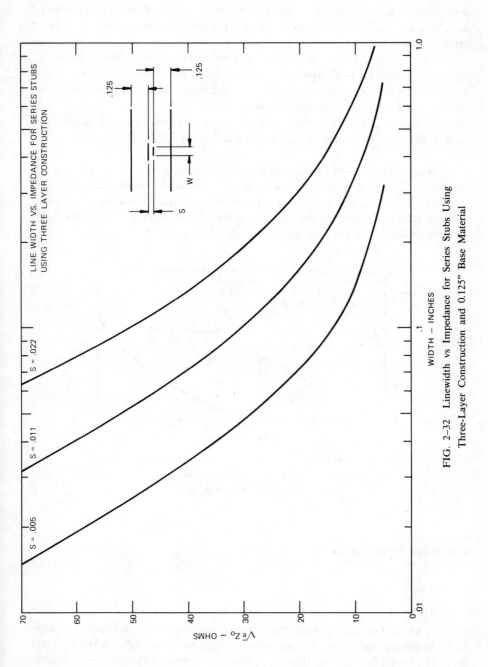

FIG. 2-32 Linewidth vs Impedance for Series Stubs Using
Three-Layer Construction and 0.125" Base Material

the one center conductor as the new ground plane, the impedance of this configuration can be shown as an extreme case of the offset line previously described. As such, its impedance can be calculated by the method described earlier in this chapter.

Figures 2–30, 2–31 and 2–32 present this data for the standard material dimensions specified in Chapter I. It will be noted that this construction does not permit as low an impedance as the first method, which will have the effect of limiting the usable bandwidth of the dc block. It should also be noted that this structure is not a true series stub, and as such, cannot be used in circuits such as filter circuits where it is necessary to have a fully shielded stub.

Despite the theoretical inaccuracy of the model, it does have remarkably predictable frequency response as a dc block series stub, provided that the line width of the coupled arm does not approach the line width of the main line, which would permit it to fringe to the opposite ground plane. Figure 2–33 shows data correlation for a 30 ohm stub built using this configuration.

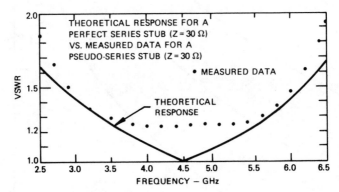

FIG. 2–33   Measured Response for the Pseudo Series Stub Compared
to the Theoretical Response for
a Perfect Series Stub

Capacitors As dc Blocks

A lumped element capacitor may also be employed as a dc block. The key to its successful use, however, depends on its installation and construction. It must be physically small so that it appears as a lumped element rather than a structure which represents a significant portion of the wavelength. It must also have short leads (ribbon leads are best) so that its stray inductance does not present problems, and be made using dielectric materials which do not introduce excessive losses at microwave frequencies. Ceramic chip capacitors with ribbon leads have been used successfully by a number of people, and at this writing, they are probably the most effective device when a lumped capacitor is desired, although MOS capacitors are rapidly coming into use in stripline, as well as microstrip, for this purpose.

Figure 2–34 shows the frequency response over the 1.0 to 14.0 GHz range for several lumped capacitor dc blocks in a 50 ohm line.

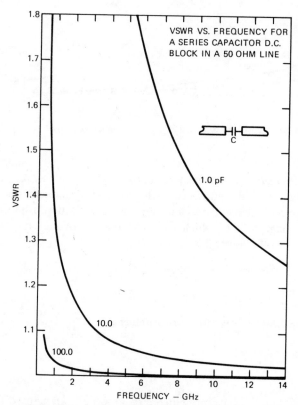

FIG. 2–34   VSWR vs Frequency for a Series Capacitor
D. C. Block in a 50 Ohm Line

Similarly, the low frequency response for a number of capacitors in the frequency range of 10 to 1000 MHz is shown in Figure 2–35.

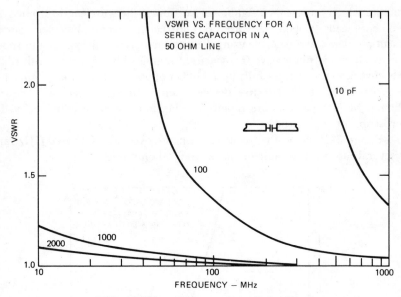

FIG. 2–35   VSWR vs Frequency for a Series Capacitor
in a 50 Ohm Line

It should be recognized that the responses presented are for theoretically perfect capacitors and do not take into account losses or stray inductances associated with realizable circuit elements. As a result, some degradation from these curves will occur, the magnitude of which will be a function of the type of capacitor chosen and the installation techniques used.

Combination dc Block/ dc Return Structures

Several authors[12][13] have presented circuits which can be used for both dc blocking and dc returns; or for bias injection points. These combination circuits make use of the interaction of series and shunt stubs to permit broad band operation and consist typically of a quarter-wavelength-long series stub spaced from a quarter-wavelength shunt stub by a quarter-wave section of line at the characteristic impedance. This structure has a two-pole VSWR response similar to that shown in Figure 2–24. The circuit is more suited to coaxial line construction where a greater range of impedances is achievable, particularly higher impedances. However, they can be built in stripline provided that a multilayer construction is used. It is imperative that a true series stub be employed. The inner conductor of the series stub must be fully shielded from the ground planes as shown in Figure 2–28a. The shunt stub may be either grounded for a dc return or, as suggested by Mouw,[13] it may be bypassed by a large capacitor to permit its use as a bias input, dc monitor, or IF output. Figure 2–36 is a design curve for this structure for the ranges of impedance available in stripline.

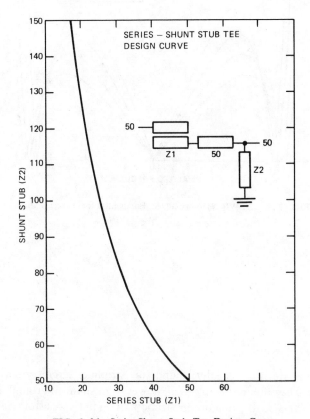

FIG. 2–36   Series-Shunt Stub Tee Design Curve

It provides the value of the optimum series stub impedance Z_1, given the available high impedance shunt stub value Z_2. It should be noted that the higher the value of Z_2, the greater the bandwidth and the lower maximum VSWR. This can be seen from Figure 2–37, which shows four typical response curves for this design method.

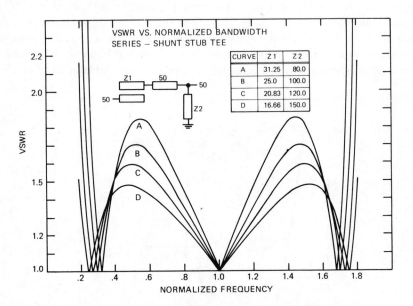

FIG. 2–37   VSWR vs Normalized Bandwidth for a Series-Shunt
Stub Tee

Resistors

Resistors are not usually found in stripline circuitry in a series configuration. They are frequently used, however, in shunt to ground as 50 ohm terminations. This is the major application of lumped resistors in stripline circuitry, and resistors of this type are available from a number of manufacturers in various package styles, ranging from 50 ohm carbon resistors which are usable in low frequencies below 1000 MHz, to the more complex packaged Pill type 50 ohm terminations, which are usable up through the higher frequency ranges.

One additional application of resistors in stripline circuit work is the packaged "Tee" pad. This consists of three resistors in a series-shunt configuration as shown in Figure 2–38.

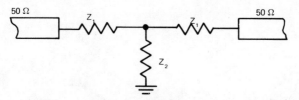

dB	Z_1	Z_2	dB	Z_1	Z_2
1.0	2.875	433.3	16.0	36.32	16.26
2.0	5.731	215.2	17.0	37.62	14.41
3.0	8.549	141.9	18.0	38.82	12.79
4.0	11.31	104.8	19.0	39.91	11.36
5.0	14.00	82.24	20.0	40.91	10.1
6.0	16.61	66.93	23.0	43.39	7.115
7.0	19.12	55.80	25.0	44.67	5.641
8.0	21.53	47.31	26.0	45.23	5.024
9.0	23.81	40.59	27.0	45.72	4.476
10.0	25.97	35.14	30.0	46.93	3.165
11.0	28.01	30.62	35.0	48.25	1.778
12.0	29.92	26.81	40.0	49.00	1.000
13.0	31.71	23.57	45.0	49.44	0.562
14.0	33.37	20.78	50.0	49.68	0.316
15.0	34.90	18.36	55.0	49.82	0.177

FIG. 2–38   Design Values for Resistive "Tee" Pads

By proper choice of values a 50 ohm match can be maintained over a wide range of attenuation. The device is frequency-insensitive if the resistors are perfect, and therefore, the only frequency limitation is the physical quality of the resistor and the method of mounting. Although commercial "Tee" pads consisting of small substrates with three resistors deposited and with metal interconnecting lands can be purchased, the user may find it desirable to manufacture his own "Tee" pad using discrete resistors. For this purpose, the chart in Figure 2–38 may be of help. It gives the value for the series resistor and the shunt resistor Z_1 and Z_2.

For high frequency applications, and more particularly for high attenuation applications, it is frequently desirable to split up the shunt resistor into two, rather than one resistor, in order to maintain line symmetry. Naturally, each of these resistors will be twice the value of the shunt resistor given in the table.

Figure 2–39 is a photograph of a number of packaged "Tee" pad elements as well as discrete resistors, capacitors and inductors discussed previously in this chapter.

FIG. 2–39   Discrete Lumped Elements for Stripline Application

Quarter-Wave Impedance Transformers

There are many occasions when it is desirable to transform the impedance of a line from its fundamental impedance to either a higher or lower level impedance, either for use as a power divider or combiner, or perhaps to enhance the aspect ratio of components which are constructed at the higher frequencies. Because of the limited range of impedances available with stripline construction, it is also frequently necessary to transpose impedance in order to provide reasonable ranges of stripwidth in the construction of a specific circuit.

The most convenient way of doing this is with the single- or multi-section quarter-wave stepped impedance transformer. In its simplest form for narrow band operation, this consists of a quarter-wave section whose impedance is the square root of the ratio of the impedance transformation. For example, a 2:1

transformer from 50 to 100 ohms would have a ratio of 2 to 1; $\sqrt{2}$ = 1.414, thus the quarter-wave transformer section between the 50 ohm line and the 100 ohm line would have an impedance of 70.7 ohms. For broader bandwidth operation, multiple section impedance transformers may be used. The response of several examples of transformer sections are shown in Figure 2–40 for single section, as well as two-, three-,and four-section transformers for a typical bandwidth of 120% and an impedance ratio of 2:1 where percentage bandwidth is defined as in equation 2–15.

$$\% \, BW = 200 \left[\frac{F_h - F_l}{F_h + F_l} \right] \qquad (2\text{-}15)$$

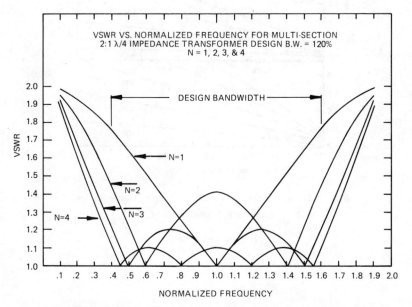

FIG. 2–40   VSWR vs Frequency for Several Impedance Transformers
of Varying Sections

Exact spacing of nulls and ripples as well as the amplitude of the ripple will vary with the impedance ratio, the number of sections and the design bandwidth. However, the fundamental responses will be consistent with Figure 2–40. The maximum VSWR of the ripple bandwidth or single section response within the design bandwidth is shown in Figure 2–41.

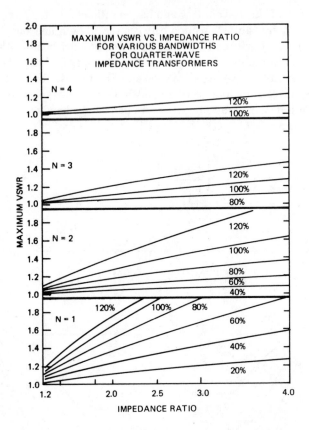

FIG. 2–41 Maximum VSWR of the Ripple Bandwidth for

Impedance Transformers of

1, 2, 3, and 4 Sections

The basic design of equal ripple impedance transformers was described by Dr. Seymour Cohn[14] in his classic paper on that subject, and was further simplified and put in tabular form in more recent publications.[15] [16] The tables presented in these references are extremely accurate for discrete impedance transformations over a wide range of impedances, as well as a multitude of sections. For solid dielectric stripline construction, the impedance ranges are normally limited to lesser values than those presented in the tables. Figures 2–42, 2–43, and 2–44 show design curves for impedance transformers over a range of 0 to 140% bandwidth for 2, 3, and 4 section transformers.

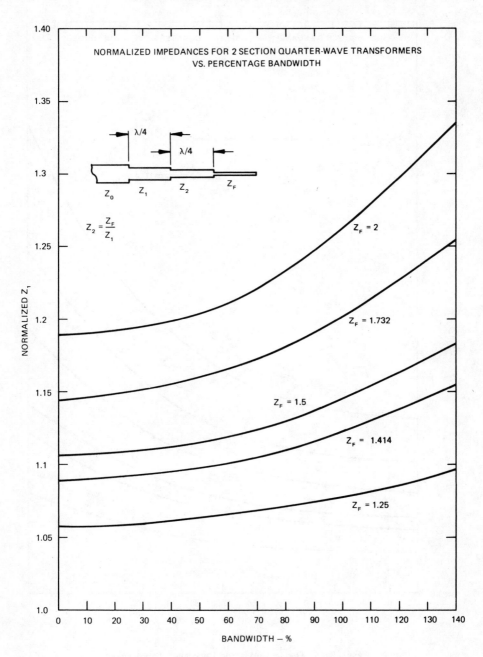

FIG. 2–42 Design Curves for Impedance Transformers
of Two Sections

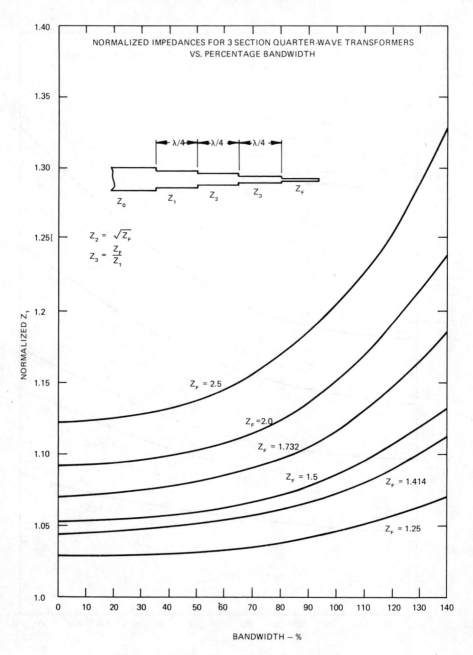

FIG. 2–43 Design Curves for Impedance Transformers
of Three Sections

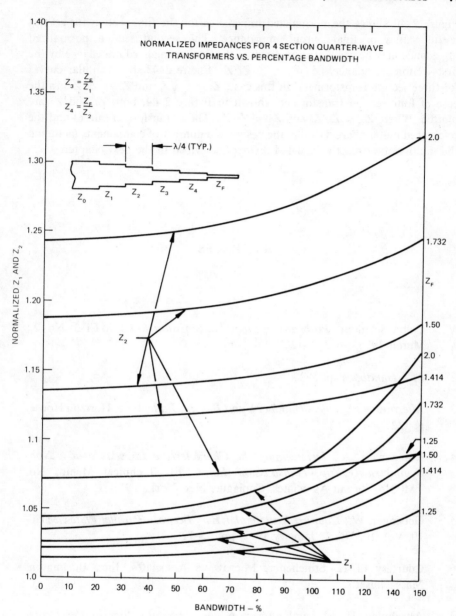

FIG. 2–44 Design Curves for Impedance Transformers
of Four Sections

Figure 2–42 shows the normalized impedance for the first section for several useful values of final output impedance. The second section normalized impedance may be calculated by dividing the final normalized impedance by the first section impedance, i.e., $Z_2 = Z_f/Z_1$. Figure 2–43 shows similar curves for three section transformers. In this case, $Z_2 = \sqrt{Z_f}$ and $Z_3 = Z_f/Z_1$. In the case of four section transformers shown in Figure 2–44, both Z_1 and Z_2 are plotted. Where $Z_3 = Z_f/Z_1$ and $Z_4 = Z_f/Z_2$. These transformer curves and the referenced material are useful in the design of a number of components including the multi-section reactive isolated dividers described in the next chapter.

REFERENCES

1   Cohn, Seymour, *Problems in Strip Transmission Lines,* MTT-3, No. 2., March 1955, pp. 119–126.

2   *ITT Handbook,* p. 592.

3   *Microwave Engineers Handbook and Buyers Guide,* 1966, Horizon House, p. 122.

4   Rosenzweig, A., *Determining the Characteristic Impedance of a Non-Symmetrical Strip Transmission Line,* LEL Technical Memo, No. TM0014, Varian Associates, Copiague, New York.

5   Getsinger, W., *Coupled Rectangular Bars Between Parallel Plates,* MTT-10, Vol. 10, No. 1, January 1962, pp. 65–72.

6   Courtesy of H. Stinehelfer, Microwave Associates, Inc., Burlington, Massachusetts.

7   Altschuler, H. M., and Oliner, A. A., *Discontinuities in the Center Conductor of Symmetric Strip Transmission Line,* MTT-8, May 1960, pp. 328–339.

8   Peters, R. W., et al, *Handbook of Tri-Plate Microwave Components*, Sanders Associates, Nashua, New Hampshire.

9   Matthaei, Young and Jones, *Microwave Filters, Impedance Matching Networks and Coupling Structures*, McGraw-Hill, New York, 1964, p. 206.

10   Adams, D. K., and Weir, W. B., *Wideband Multiplexers Using Directional Filters*, Microwaves, Vol. 8, No. 5, May 1969, pp. 44–50.

11   Young, L., *Microwave Engineers Handbook and Buyers Guide*, 1966, Horizon House, p. 106.

12   McDermott, M., and Levy, R., *Very Broadband Coaxial dc Returns Derived by Microwave Filter Synthesis*, Microwave Journal, Volume 8, No. 2., February 1965, pp. 33–36.

13   Mouw, R. B., *Broadband dc Isolator-Monitors*, Microwave Journal, Vol. 7, No. 11, November 1964, pp. 75–77.

14   Cohn, Seymour, *Optimum Design of Step Transmission Line Transformers*, MTT-3, No. 3, April 1955, pp. 16–21.

15   Young, L., *Tables for Cascaded Homogeneous Quarter-Wave Transformers*, MTT-7, No. 2, April 1959, pp. 233–237.

16   Matthaei, Young, and Jones, *Microwave Filters, Impedance Matching Networks and Coupling Structures*, McGraw-Hill, New York, 1964, pp. 275–278.

Direct Coupled Hybrids, Power Dividers, and Directional Couplers

One of the simplest and least expensive methods of making directional couplers and power dividers is by the direct coupled or branch line type of structure. This method of construction is particularly suitable to a single plane configuration and has the advantage of maintaining dc continuity. Although originally branch guide couplers and hybrids were characterized by narrow band operations, broadband devices are possible by the use of multi-section construction. This is particularly applicable in the area of in-line power dividers which are now capable of operation over multi-octave and even decade bandwidths. Broadband power dividers of this class will be discussed later in this chapter.

The most fundamental direct coupled structure is the two-section branch guide coupler. This is shown in schematic form in Figure 3–1. It consists of a main line which is coupled to a secondary line by two quarter-wavelength long sections spaced one quarter-wavelength apart, thus creating a square or circle approximately one wavelength in circumference. The coupling factor is determined by the ratio of the impedance of the shunt and series arms which also must be adjusted to maintain a proper match over the band. Because this structure is ideally suited to coupling values in the region of 3.0 to 9.0 dB, significant main arm insertion losses result because of the amount of power coupled to the secondary arm. This is shown in Figure 3–2.

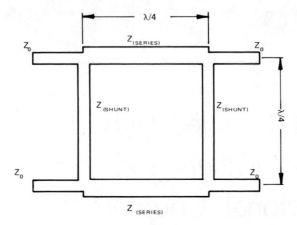

FIG. 3–1   Branch Line Configuration

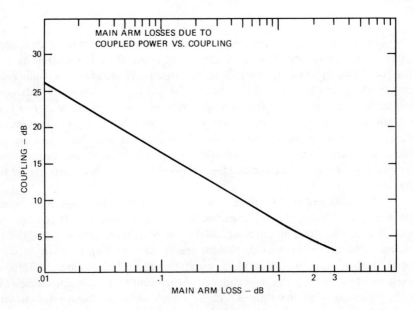

FIG. 3–2   Main-Arm Coupler Losses Due to
Coupled Power As a Function of Coupling

It should be remembered that these main arm losses are theoretical losses due to the power coupled to the secondary arm and do not include power dissipated as the result of copper and dielectric losses. The impedance ratios necessary for proper coupling, shown in equation 3–1', are expressed in terms of characteristic admittance,

$$\text{Coupling (dB)} = 10 \log_{10} \left[\frac{1 + \dfrac{Y^4 + Y^6}{Y^2}}{Y^2} \right] \qquad (3–1)$$

where characteristic admittance equals the reciprocal of characteristic impedance, i.e., $Y_o = 1/Z_o$. The relationship in equation 3–1, however, does not establish the entire requirement for a branched guide coupler, inasmuch as the junction must be properly matched. Therefore, the series quarter-wave section between the two shunt quarter-wave arms must be lowered in impedance in order to compensate for the loading of the coupling arms.

Figure 3–3 is a plot of the series and shunt arm impedances for a two-section branch-guide coupler of coupling values varying from 3.0 to 8.0 dB. Higher values of coupling have not been presented because of the inherent high impedance line limitation of solid dielectric stripline.

The frequency response of this type of coupler is shown in Figure 3–4 for values of 5.0, 6.0 and 7.0 dB, which have been chosen as typical cases. It can be observed that both the coupling and the VSWR values are reasonably flat and within tolerable limits over a 20% bandwidth. However, the directivity drops below 20 dB at greater than a 10% bandwidth, and, in fact, the limitation in directivity due to the basic structure is greater than would normally be expected as a result of main-line and secondary-arm VSWR.

3.0 dB Hybrid Case

Normally, the tightest coupling which would be desirable in such a structure is 3.0 dB, or the equal-power split case. For this construction, the shunt

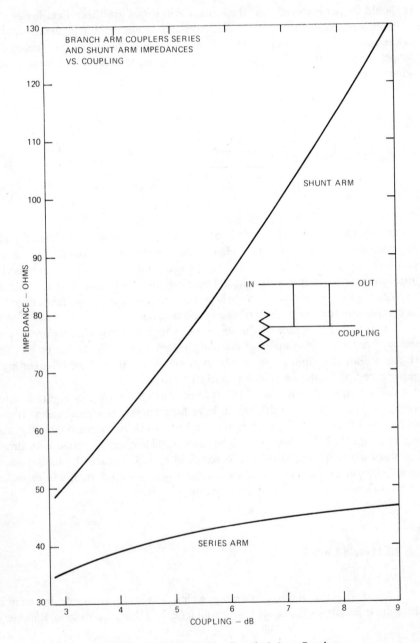

FIG. 3–3   Design Curves for Branched Arm Couplers

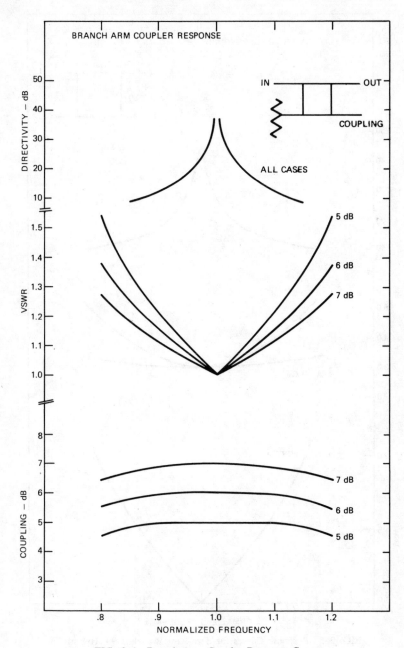

FIG. 3–4   Branch-Arm Coupler Response Curves

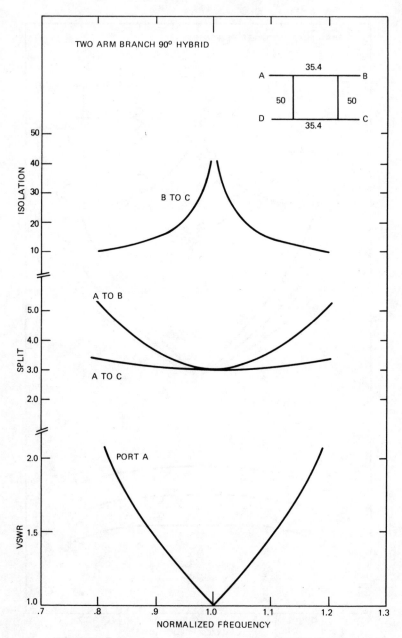

FIG. 3–5   Response Curves for Two-Arm Branched 90° Hybrids

arms become equal to the characteristic impedance of the input and output arms, or in our normal case, 50 ohms. The series arms are set at 35.4 ohms which is equal to the input arm and shunt arm impedances divided by the square root of two. The response of this structure is shown in Figure 3–5. As in the case of the looser coupled branch-line structure, the VSWR and coupling split are acceptable over approximately a 20% bandwidth, while the isolation is usable only over approximately a 10% bandwidth. Additionally, Figure 3–6 shows the phase response for this circuit.

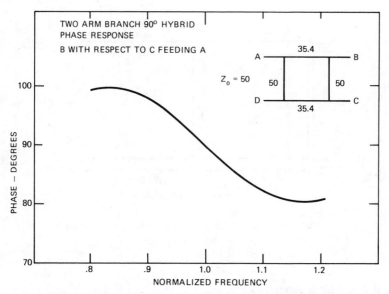

FIG. 3–6   Phase Response for a Two Arm Branched 90° Hybrid

This type of hybrid is fundamentally a 90° hybrid; that is, the two output arms which are equal in amplitude are in a 90° phase relationship to each other. This 90° relationship is perfect only at the design frequency and varies according to Figure 3–6 with frequency about the ideal design frequency. Although this 90° relationship is frequency sensitive, it varies only approximately $\pm 5°$ over a 10% bandwidth, making it usable for many applications when narrow bands are acceptable; for example, power division for image rejection mixers and single sideband modulators as well as circuits requiring reflection of mismatches into a terminated fourth port load.

The fundamental limitation of the circuit element is, of course, its bandwidth, which is narrow. This restriction can be overcome by adding additional sections which, in theory, is an acceptable technique for broad banding. In practice, this is possible for coaxial structures where a wider range of impedances is possible. In stripline, however, it is difficult to achieve more than a three-section coupler, inasmuch as the end sections require impedances which reach the upper limits of practical realization. Several three-section designs are possible; however, the broadest band version of these is described by Reed and Wheeler[2], and consists of the circuit shown in Figure 3–7.

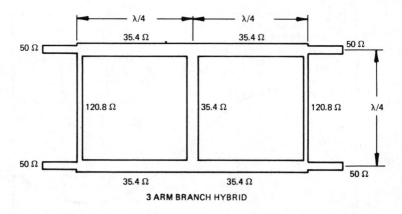

FIG. 3–7   Design Parameters for a Three-Arm Branch
90° Hybrid

The end coupling sections have impedances of 120.8 ohms which is about the upper limit of practical construction for most stripline configurations. Over a 25% bandwidth, this circuit has a maximum theoretical VSWR of 1.12 with a theoretical isolation of 25.3 dB and a maximum unbalance of 0.5 dB. A similar configuration in which all of the series arms are equal to the characteristic impedance, while the end shunt arms are equal to 120.8 ohms, and the center shunt arm is equal to 70.7 ohms, is useful with higher frequencies where the wider linewidths required by the lower impedance may create an undesirable aspect ratio, due to the shortened quarter-wave sections. Its performance is not as favorable inasmuch as the maximum VSWR for the same bandwidth is 1.2, the isolation is 20.5, and the unbalance is 0.6 dB. In general, the multi-section branched-line coupler is not a desirable circuit for stripline applications, since better performance can be achieved in much less space by means of a quarter-wave coupled line hybrid which will be discussed in later chapters. The branched-guide device, while having the advantage of dc coupling, has the disadvantage of requiring a greater amount of circuit area, increased line lengths which result in additional insertion losses, and restricted bandwidths over the single quarter-wave section and multiple quarter-wave section devices in common usage.

Magic Tees

While the previously described branch arm coupler has a 90° phase response, the most common 0–180° phase response device is the so-called rat-race or 1.5 wavelength circumference ring. This is illustrated in Figure 3–8.

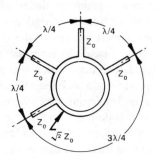

FIG. 3–8   Design Parameters for a 1.5 Wavelength
Rat-Race Magic Tee

It consists of a ring 1.5 wavelengths in circumference having four arms separated by 60° of angular rotation. Thus for a common input arm, there are two output arms spaced one quarter-wavelength away and a fourth terminated arm spaced a quarter-wavelength away from one of the output arms and three quarters of a wavelength away from the other output arm. At center frequency, the output split from the common input arms to the two output arms is equal; however, the phase relationship between them is 180°. Similarly, if power is fed in the fourth arm, the power split to the two previously mentioned output arms is also equal; however, the phase relationship between them is 180°. The design of the device requires that the impedance of the central ring be equal to the characteristic impedance times the square root of two, which, for a normal 50 ohm system, means that this main ring impedance is equal to 70.7 ohms. The response of this circuit, both in terms of its isolation, split, and VSWR is shown in Figure 3–9. As can be seen from this curve, the useful bandwidth in terms of isolation, split, and VSWR is approximately 20% depending upon the specific circuit requirements. Figure 3–10 shows the phase relationship for both 0 and 180° conditions for this type of structure. The 0° condition has a far greater bandwidth for the same phase deviation than does the 180° phase relationship. Nevertheless, the performance is adequate for most circuit configurations operating in bandwidths up to 20%. This is true for balanced mixers, image rejection mixers, single sideband generators, monopulse comparators, and other circuits requiring a depth null, or signal cancellation in the order of 20 dB.

There are alternate circuit versions of the magic tee rat-race in which greater sections longer than one quarter-wavelength are used as the fundamental element of construction. The simplest of these is the three quarter/five quarter wavelength rat-race hybrid. In this case, the main section is three quarters of a wavelength long and the extended section, which was previously three quarters of a wavelength is now five quarters of a wavelength long. This class of extended wavelength circuit is useful at higher frequencies where it is difficult to maintain a proper aspect ratio or where it is desirable to fold the arms inward on the ring and provide additional circuit elements internal to the hybrid circle, as in the case, for example, of the balanced mixer. It carries the distinct disadvantage of a narrowed bandwidth due to the increased length of the individual sections. This can be seen in Figure 3–11 which shows the VSWR, power split and isolation of this ring. Comparison of Figure 3–11 with 3–9 will show a bandwidth compression of approximately 50% for the performance operation. Figure 3–12 shows the phase characteristics for this circuit and illustrates a similar compression of performance bandwidth when compared with Figure 3–10. The rat-race form of hybrid may also be constructed as an unequal split device as shown in Figure 3–13a.[3][4]

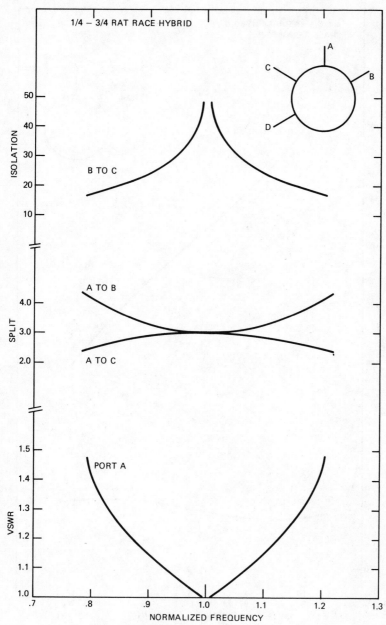

FIG. 3–9   Response Curves for a 1.5 Wavelength
Rat-Race Magic Tee

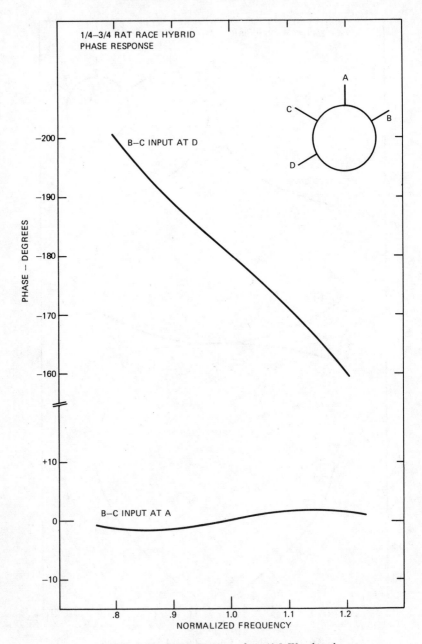

FIG. 3–10 Phase Response for a 1.5 Wavelength
Rat Race Magic Tee

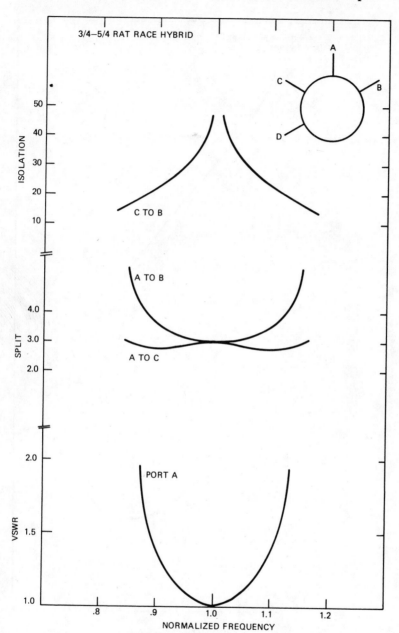

FIG. 3–11   Response Curves for a 3.5 Wavelength
Rat-Race Magic Tee

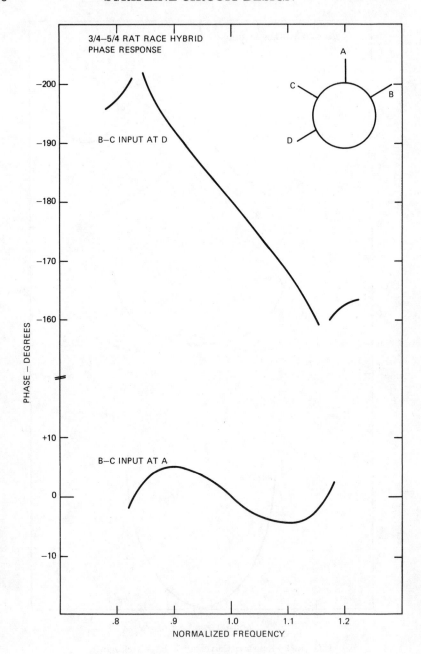

FIG. 3–12   Phase Response for a 3.5 Wavelength
Rat Race Magic Tee

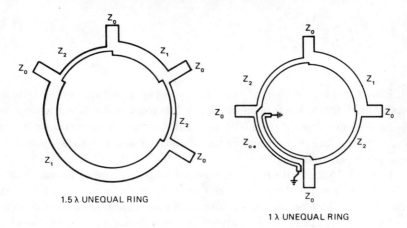

1.5 λ UNEQUAL RING

1 λ UNEQUAL RING

FIG. 3–13   Impedance Definitions for Two Types of
Unequal Split Ring Couplers

Each section is a quarter-wavelength long as in the equal split tee, except for the line between port one and port four, which is three quarters of a wavelength long. As in the case of the branched guide coupler, the power split between the arms is a function of the characteristic admittance of the ring sections where adjacent sections have unequal admittances and the opposite sections are equal. The characteristic admittances of the ring sections may be calculated from equations 3.2 and 3.3, keeping in mind that

$$\frac{Y_1}{Y_0} = \left(\frac{P_3}{P_2} \right)^{\frac{1}{2}} \tag{3-2}$$

$$\frac{Y_2}{Y_0} = \left(\frac{P_4}{P_2} \right)^{\frac{1}{2}} \tag{3-3}$$

the condition of equation 3–5 must be met in order to provide a perfect match at the design frequency.

$$Y_1{}^2 + Y_2{}^2 = Y_0{}^2 \tag{3-4}$$

Figure 3–14 is a plot of the ring impedances versus coupling factor expressed in dB for the range of impedances achievable in solid dielectric stripline. It should be noted that for the 3.0 dB or equal split case, both curves for Z_1 and Z_2 converge on 70.7 ohms which is the established design for the normal 1.5 wavelength equal split ring.

Figure 3–15 shows the response curve for a 6.0 dB coupling unequal hybrid ring. As in the case of the equal split ring, normal performance is limited to approximately a 20% or 30% bandwidth maximum, depending upon the degree of performance required. The bandwidth of this device may be increased to approximately an octave by substituting the coupled line quarter-wave section for the three quarter-wave section of the original 1.5 lambda ring. This circuit would then make a ring equal to 1.0 lambda in circumference and is shown in Figure 3–13(b). It is necessary to short-circuit the ends of the coupled line sections, which makes this difficult to construct and, therefore, limits its use to the lower frequencies where these short circuits may be conveniently constructed. The characteristic impedance of the coupled line section may be described in terms of the even- and odd-mode impedances, Z_{oe} and Z_{oo}. In order to make use of the design curves provided in Chapter IV, only the normalized Z_{oe} is necessary and this may be calculated from equation 3–5.

$$Y_{oe} = \left(\sqrt{2} - 1\right) Y_1 = \frac{1}{Z_{oe}} \tag{3-5}$$

An additional broadband magic tee concept was developed by Alford and Watts[5] and was further developed in stripline form by Tatsuguchi[6]. These hybrids have not found wide acceptance because of their extreme difficulty of construction, although it has been the author's experience that at frequencies through S-band, their performance is excellent when carefully manufactured.

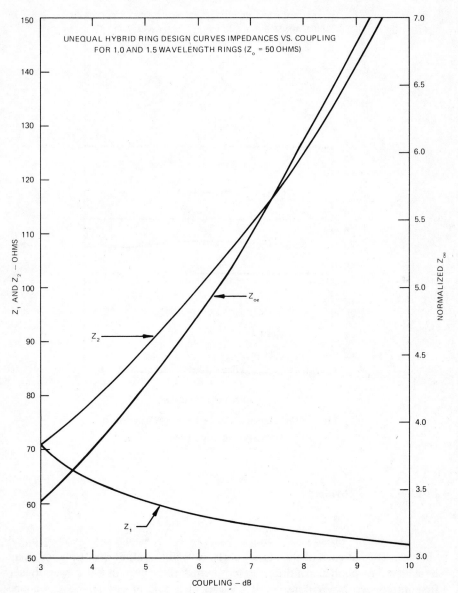

FIG. 3–14 Design Curves for Unequal Split Ring
Couplers

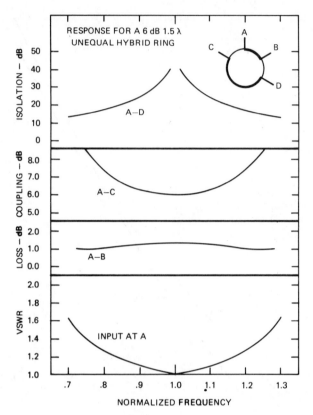

FIG. 3–15   Response Curves for a 1.5 Wavelength
Unequal Split Ring Coupler

In-Line Power Dividers

A useful broadband circuit for power division with equal phase characteristics at each of the output ports, as well as output port isolation, may be achieved through use of the series terminated, three port, in-line power divider first introduced by Wilkinson[7]. It consists of a pair of quarter-wave sections having a characteristic impedance of 70.7 ohms which are series terminated at the output with a 100 ohm resistor, thus providing essentially three port circuit. The schematic for this device is shown in Figure 3–16(a).

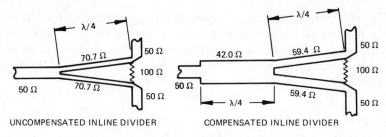

FIG. 3–16   Compensated and Uncompensated In-Line
Three Port Power Dividers

Its performance, although excellent, may be improved by the addition of a
quarter-wave transformer in front of the power division step and a shift in the
impedance levels as shown in Figure 3–16(b)[8].

The usable bandwidth of this circuit is approximately one octave and its
performance for both the compensated and uncompensated case is shown in
Figure 3–17.

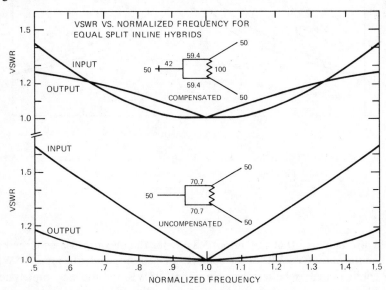

FIG. 3–17   VSWR vs Normalized Frequency for Equal
Split In-Line Hybrids of the Compensated
And Uncompensated Types

The power division accuracy is not frequency sensitive and is, therefore, strictly a function of the construction of the device. At the band edges, where isolation is not as high as at the design frequency, it may also be affected by the load impedances. This isolation characteristic is shown in Figure 3–18, and can be seen to be slightly improved for the compensated case.

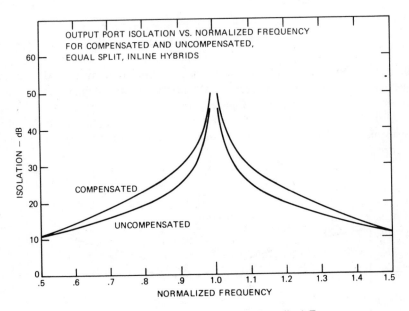

FIG. 3–18 Co-linear Arm Isolation vs Normalized Frequency
for Compensated and Uncompensated Equal
Split In-Line Hybrids

It should also be noted that the output VSWR for the uncompensated circuit is better than the output VSWR for the compensated circuit, while the input VSWR for the compensated is better than the input VSWR for the uncompensated circuit. Therefore, the choice of circuit may be dictated by the input/output VSWR characteristic as well as by whether or not the necessary space exists to add the extra quarter-wave transformer.

Unequal Split In-Line Power Dividers

The unequal split case of the in-line power divider is illustrated in the schematic of Figure 3–19.

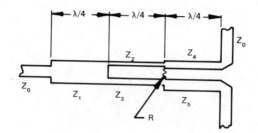

FIG. 3–19   Circuit Configuration for an Unequal Split
In-Line Three Port Coupler

As can be seen, it is necessary to add output impedance transformers to the normal three port circuit. For both the case without the input impedance transformer, and for the improved or compensated case which includes the input impedance transformer, design equations have been developed and are shown in equation groups 3–6a through 3–6f, and 3–7a through 3–7g.

$$\frac{1}{K^2} = \frac{P_a}{P_b} \tag{3–6a}$$

$$Z_2 = Z_0 \sqrt{K(1 + K^2)} \tag{3–6b}$$

$$Z_3 = Z_0 \sqrt{\frac{1 + K^2}{K^3}} \tag{3–6c}$$

$$Z_4 = Z_0 \sqrt{K} \tag{3–6d}$$

$$Z_5 = \frac{Z_0}{\sqrt{K}} \tag{3-6e}$$

$$R = Z_0 \frac{1 + K^2}{K} \tag{3-6f}$$

$$K^2 = \frac{P_b}{P_a} \tag{3-7a}$$

$$Z_1 = Z_0 \left[\frac{K}{1 + K^2} \right]^{\frac{1}{4}} \tag{3-7b}$$

$$Z_2 = Z_0 \left[K^{\frac{3}{4}} \ (1 + K^2)^{\frac{1}{4}} \right] \tag{3-7c}$$

$$Z_3 = Z_0 \left[\frac{(1 + K^2)^{\frac{1}{4}}}{K^{5/4}} \right] \tag{3-7d}$$

$$Z_4 = Z_0 \ \sqrt{K} \tag{3-7e}$$

$$Z_5 = \frac{Z_0}{\sqrt{K}} \tag{3-7f}$$

$$R = Z_0 \left[\frac{(1 + K^2)}{K} \right] \tag{3-7g}$$

This latter equation group, 3–7a through 3–7g, is presented graphically in Figure 3–20 for a variety of coupling values from 3.0 to 10.0 dB.

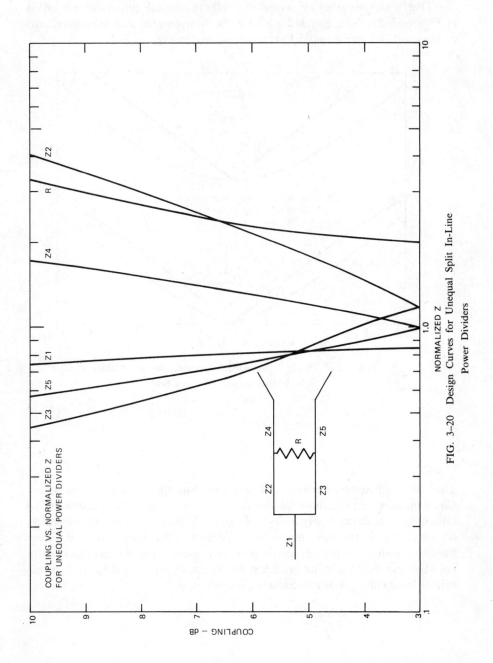

FIG. 3–20   Design Curves for Unequal Split In-Line
Power Dividers

Performance curves for several examples of unequal splitters are presented in Figures 3–21, 3–22 and 3–23. VSWR for compensated and uncompensated designs with 2:1 ratios and 3:1 ratios are shown in Figures 3–21 and 3–22.

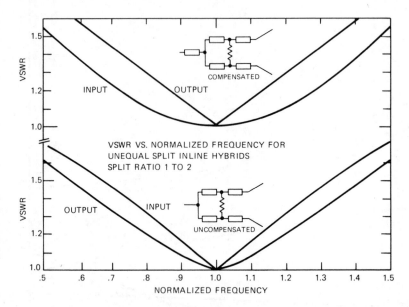

FIG. 3–21 VSWR vs Normalized Frequency for an Unequal Split In-Line Hybrid Having a Split Ratio Of 2:1

The bandwidth response narrows as the ratio of unequal power split increases. Coupling, however, remains relatively flat across the bandwidth regardless of the ratio of split, as shown in Figure 3–23. In general, this circuit element, regardless of whether it is the uncompensated or compensated design, is useful over bandwidths up to one octave, and therefore, represents a major step in bandwidth improvement over the other branched line circuits previously described, which exhibit bandwidths of approximately 20–30% at most.

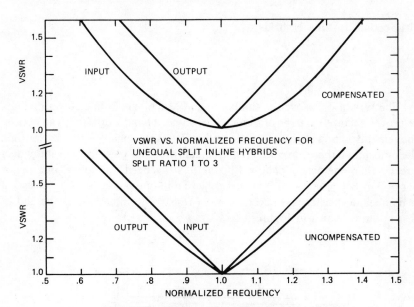

FIG. 3-22   VSWR vs Normalized Frequency for
an Unequal Split In-Line Hybrid Having
a Split Ratio Of 3:1

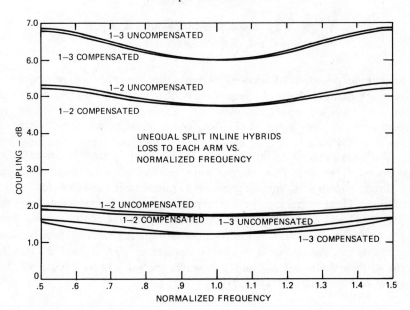

FIG. 3-23   Coupling vs Frequency for Several
Types of Unequal In-Line Power Dividers

Multi-Section, In-Line Hybrids

The Wilkinson, as well as the Parad and Moynihan, in-line power dividers have proven themselves to be of tremendous value at bandwidths up to an octave. Although this bandwidth was considered to be exceptionally wide at the time of their inception, there have been requirements in recent years for power dividers of greater than one octave bandwidth. In 1968 this need was solved in a major paper by Dr. Seymour Cohn[10]. He described an in-line power divider consisting of a number of quarter-wave sections with resistive terminations at the end of each section as shown in Figure 3–24.

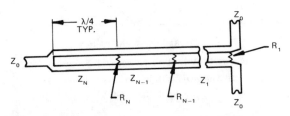

FIG. 3–24   Circuit Configuration for a Multiple Section In-Line Equal Split Power Divider

As the number of individual sections increases, the bandwidth of the device and its inherent isolation between output ports also increases. By means of this technique, improved operation over one octave, as well as multi-octave, performance has been achieved. In fact, power dividers with bandwidths of greater than a decade have been constructed with remarkable performance.

The design procedure for these devices is relatively straightforward. The impedances of the individual quarter-wave sections are established from the normalized impedances for quarter-wave transformer sections for a 2:1 transformer. These values may be determined from the many tables, or from the figures, 42, 43 and 44 in Chapter II. Once the impedances of the sections have been determined, the values of the resistors terminating each of these sections may also be calculated. For a two-section power divider, R_1 and R_2 may be calculated from equations 3–8 and 3–9, where the value of θ_3 in equation 3–8 is determined by equation 3–10.

$$R_2 = \frac{2\,Z_1 Z_2}{\sqrt{(Z_1 + Z_2)(Z_2 - Z_1 \cot^2 \Phi_3)}} \qquad (3\text{--}8)$$

$$R_1 = \frac{2\,R_2\,(Z_1 + Z_2)}{R_2\,(Z_1 + Z_2) - 2Z_2} \qquad (3\text{--}9)$$

$$\Phi_3 = 90° \left[1 - \frac{1}{\sqrt{2}} \left(\frac{f_2/f_1 - 1}{f_2/f_1 + 1} \right) \right] \qquad (3\text{--}10)$$

These designs for bandwidths from 1.2:1 to 4:1 are summarized in Figure 3–25.

For power dividers having more than two-sections, the design equations become more complicated. It is convenient to deal in terms of conductance and admittance, rather than resistance and impedance.

$$R = \frac{1}{G} \qquad (3\text{--}11)$$

For this design the characteristic impedance of each of the transformer sections is chosen as before, and the resistances terminating each section are calculated from equations 3–12 for the output resistor; 3–13 for the intermediate resistors, making use of the relationship in 3–14; and by equation 3–15 for the first resistor terminating the input section.

$$G_1 = (1 - Y_1) \qquad (3\text{--}12)$$

$$G_K = \frac{Y_{k-1} - Y_k}{Y_{k-1} T_1 T_2 \cdots T_{k-1}} \text{, for } K = 2 \text{ to } N\text{-1} \qquad (3\text{--}13)$$

$$T_k = \frac{4Y_{k-1}Y_k}{(Y_{k-1} + Y_k + 2G_k)^2} \qquad \text{for } K = 1 \text{ to } N \qquad\qquad (3\text{--}14)$$

$$G_n = \cfrac{\frac{1}{2}Y_n^2{}_{-1}}{-2G_{n-1} + \cfrac{Y_n^2{}_{-2}}{-2G_{n-2} + \cfrac{Y_n^2{}_{-3}}{\cfrac{\vdots}{-2\ G_2 + \cfrac{Y_1^2}{-2G_1 + 1 + 0.7(S_{e',\,90^\circ} - 1)}}}}}$$

where:

$$S_{e',\,90^\circ} = 1 \text{ for N odd} \\ = S_{em}, \text{ for N even} \qquad\qquad (3\text{--}15)$$

In order to make use of equation 3–15, it is necessary to know the maximum ripple VSWR for the prototype transformer sections chosen for the design. This may be derived either from the previously mentioned tables, or from Figure 41 in Chapter II. This data has been replotted, along with the output VSWR's in Figure 3–26.

       Complete design summary for the three-section power dividers may be found in Figure 3–27, and similar data for four-section power dividers in Figure 3–28.

       The performance characteristic of this class of device is similar to that of the single section except that the VSWR and isolation characteristics follow the classic equal ripple functions. The minimum VSWR at both the input and output ports occurs at the same point as does the maximum isolation between the two output ports. This maximum isolation theoretically goes to infinity and

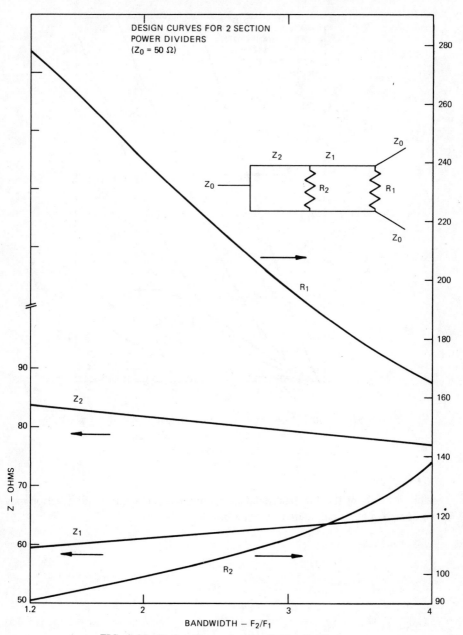

FIG. 3–25   Design Curves for Two-Section In-Line
Equal Split Power Dividers

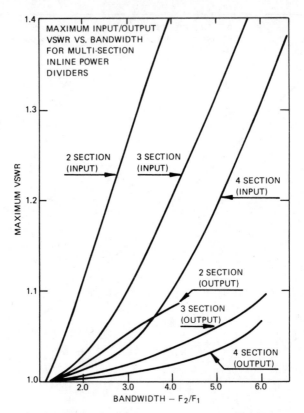

FIG. 3–26   Maximum Input-Output VSWR vs
Bandwidth for a Multi-section In-Line
Power Divider

is limited only by the constructional techniques used. The minimum isolation may
be approximately calculated by equation 3–16.

$$\text{Isol}_{(min)} \approx 20 \, \text{Log}_{10} \left(\frac{2.35}{S_{em} - 1} \right) \, dB \qquad (3\text{–}16)$$

where

$$S_{in} = S_{em} = \text{Maximum Ripple VSWR} \qquad (3\text{--}17)$$

Again, as with previously described devices, it should be recognized that the performance characteristics presented are theoretical and will be degraded by some factor which is a function of the quality of construction, both in terms of tolerances and accuracy of calculations as well as physical quality of the resistors used in the fabrication of the device. This degradation is generally small in the low frequency regions of L- through S-band; however, it can become substantial as designs approach X- and Ku-band, and in fact, designs of this type tested in the Ku-band frequency regions have been found to be as much as 35% and 40% lower in minimum isolation than the theoretical curves would suggest.

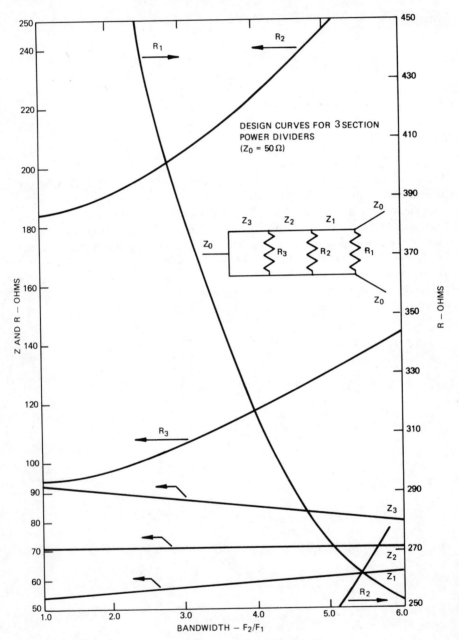

FIG. 3–27 Design Curves for Three-Section
In-line Equal-Split Power Dividers

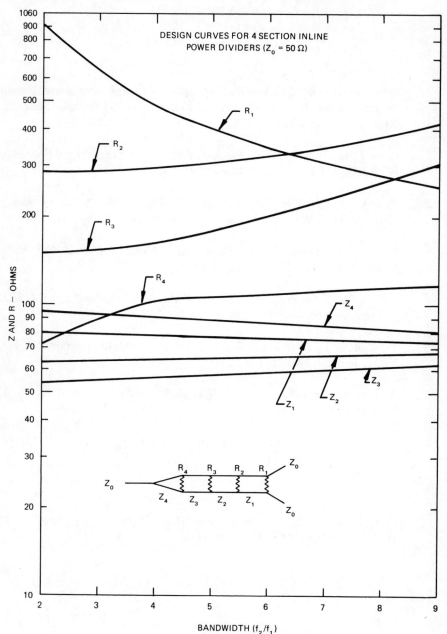

FIG. 3–28   Design Curves for Four-Section
In-line Equal-Split Power Dividers

REFERENCES

1   Sferrazza, P. J., *Unidirectional Coupler Research and Development Program,* Final Report No. 5224–1240–8, Sperry Gyroscope Company, Great Neck, New York, 1953.

2   Reed, J., and Wheeler, G. J., *Method of Analysis of Symmetrical Four-Port Networks,* MTT-4, No. 4, October 1956, pp. 246–252.

3   Pon, C. Y., *Hybrid Ring Directional Couplers for Arbitrary Power Division,* MTT-9, No. 6, November 1961, pp. 529–535.

4   Pon, C. Y., *Control Your Power Split Using a Hybrid Rat Race,* Microwaves, Vol. 9, No. 4., April 1970, pp. 34–37.

5   Alford, A., and Watts, C. B., *A Wide Band Coaxial Hybrid,* IRE Convention Record, 1956, Part 1, pp. 176–179.

6   Tatsuguchi, I., *UHF Strip Transmission Line Hybrid Junction,* MTT-9, No. 1, January 1961, pp. 3–6.

7   Wilkinson, Ernest, *An N-Way Hybrid Power Divider,* MTT-8, No. 1, January 1969, pp. 116–118.

8   Parad, L. I., and Moynihan, R. L., *Split Tee Power Divider,* MTT-13, No. 1, January 1965, pp. 91–95.

9   ibid

10   Cohn, S., *A Class of Broadband, Three-Port to TEM Mode Hybrids,* MTT-16, No. 2, February 1968, pp. 110–118.

CHAPTER

—4—

Coupled Parallel Lines

Perhaps one of the most useful and widely applied stripline structures is the quarter-wave parallel-coupled line. It forms the basis of most 90° hybrids, directional couplers and bandpass filters, as well as certain types of phase shift networks. In fact, many broadband multi-section coupled devices that are in use today would not be possible without this basic coupling structure.

In the case of directional couplers, this structure consists of one or more quarter-wavelength sections which are coupled in proximity to each other to a greater or lesser extent, depending upon the requirements of the circuitry. For bandpass filters, quarter- or half-wave resonators are coupled in a similar manner. Certain types of phase shift networks also make use of quarter-wave coupled sections either single, in multiple sections, or in tandem sections, all of which will be discussed in later chapters.

The requirements for parallel-coupled lines fall into three categories: those for loosely coupled lines, as in very narrow bandpass filters and loosely coupled directional couplers; very tight coupling, as in the case of 90° hybrids and broadband circuits; and a method of providing a wide range of coupling which covers the loose, the tight and the intermediate range conditions. Although, as the bibliography shows, many people have contributed to the literature; two key men are paramount in the discussion of parallel-coupled lines. These are Dr. Seymour Cohn who formulated the initial theory for loose and very tightly coupled line, and Paul Shelton, who covered the wide range variable case. The equations, and the resultant nomographs and design curves which have been generated from them and which are presented in this chapter are almost entirely due to the work of these two men, except as otherwise referenced. Without their significant effort, stripline and, most notably, broadband stripline, would not have advanced to the point at which it is today.

Basic Concept

As described by Cohn[1], the characteristic impedance of a pair of coupled lines can be described in even- and odd-mode characteristic impedances as illustrated by the electric field distributions in Figure 4–1.

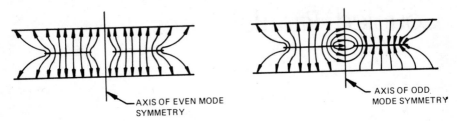

AXIS OF EVEN MODE SYMMETRY

AXIS OF ODD MODE SYMMETRY

EVEN MODE ELECTRIC FIELD DISTRIBUTION

ODD MODE ELECTRIC FIELD DISTRIBUTION

FIG. 4–1

The even-mode characteristic impedance (Z_{oe}) is the impedance measured from one strip to ground where the strips are at the same potential and have equal currents in the same direction. This is described by:

$$Z_{oe} = \frac{30\,\pi}{\sqrt{\epsilon}} \cdot \frac{K\ (ke')}{K\ (ke)}\ \text{ohms} \qquad (4\text{-}1)$$

where K is a complete elliptic integral of the first kind and

$$ke = \tanh\left(\frac{\pi}{2} \cdot \frac{w}{b}\right) \cdot \tanh\left(\frac{\pi}{2} \cdot \frac{w+s}{b}\right) \qquad (4\text{-}2)$$

$$ke' = \sqrt{1 - k_e^2} \tag{4-3}$$

where ϵ is the dielectric constant and w, b, and s are the dimensions shown in Figure 4–2a. The odd-mode characteristic impedance (Z_{oo}) is defined as that impedance from each strip to ground when the strip potentials are opposite and the currents are in opposite directions.

$$Z_{oo} = \frac{30\,\pi}{\sqrt{\epsilon}} \cdot \frac{K\ (ko')}{K\ (ko)} \tag{4-4}$$

where

$$k_0 = \tanh\left(\frac{\pi}{2} \cdot \frac{w}{b}\right) \cdot \coth\left(\frac{w}{2} \cdot \frac{w+s}{b}\right) \tag{4-5}$$

$$k_0' = \sqrt{1 - k_0^2} \tag{4-6}$$

These relations show that the characteristic impedances will be different for each mode and that the even mode will always be higher. Because of the difficulty of evaluating the complete elliptic integrals, it is frequently desirable to express Z_{oe} and Z_{oo} in terms of the fringing capacities, which is applicable for the zero thickness strip case. This is shown in equations 4–7 and 4–8.

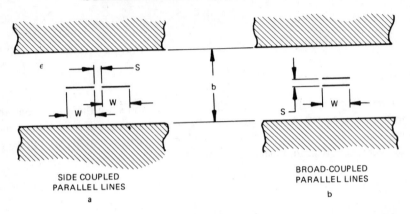

SIDE COUPLED
PARALLEL LINES
a

BROAD-COUPLED
PARALLEL LINES
b

FIG. 4-2

$$Z_{oe} = \frac{94.15}{\sqrt{\epsilon} \left[\dfrac{w}{b} + \dfrac{1}{2\epsilon} \left(C_f'(0) + C_{fe}'(0), \dfrac{s}{b} \right) \right]} \text{ ohms} \tag{4-7}$$

$$Z_{oo} = \frac{94.15}{\sqrt{\epsilon} \left[\dfrac{w}{2\epsilon} + \dfrac{1}{2\epsilon} \left(C_f'(0) + C_{fo}' \ 0, \dfrac{s}{b} \right) \right]} \text{ ohms} \tag{4-8}$$

This will hold good accuracy where w/b is greater than or equal to 0.35. For this case, the relations can be expressed in a more usable form with a resultant accuracy of about 1%. This applies to the usual solid dielectric strip case, where the strip thicknesses are not exactly zero, but are very close to zero in terms of the overall ground plane spacing and normal strip thickness.

$$\frac{C_f'(0)}{\epsilon} = \frac{2}{\pi} \log_e 2 = 0.4407 \tag{4-9}$$

$$\frac{C_{fe}'\left(0, \dfrac{s}{b}\right)}{\epsilon} = \frac{2}{\pi} \log_e \left(1 + \tanh\left(\dfrac{\pi s}{2b} \right) \right) \tag{4-10}$$

$$\frac{C_{fo}'\left(0, \frac{s}{b}\right)}{\epsilon} = \frac{2}{\pi} \log_e\left(1 + \coth\left(\frac{\pi s}{2b}\right)\right)$$

(4-11)

Design Curves for Side-Coupled Lines

In practice, of course, it is necessary to relate the even- and odd-mode impedances to the desired value of coupling between the parallel-coupled lines. This is shown by:

$$C_{(dB)} = -20 \log_{10}(C_0)$$

(4-12)

Where C_0 = Voltage Coupling Ratio

$$\text{Norm. } Z_{oe} = \sqrt{\frac{1 + C_0}{1 - C_0}}$$

(4-13)

$$\text{Norm. } Z_{00} = \sqrt{\frac{1 - C_0}{1 + C_0}}$$

(4-14)

Thus, it can be seen that both the even- and odd-mode characteristic impedances relate to the voltage coupling ratio and, in fact, for a matched condition within the quarter-wave section, equation 4–15 must be satisfied.

$$Z_0^2 = Z_{oe}\ Z_{00}$$

(4-15)

It should be obvious from the above that for the matched case, which is of course the normal condition, the coupling ratio, as well as the dimensions as a function of the construction, can be defined by the even-mode impedance alone. Therefore, throughout the remainder of this book, for brevity and simplicity, the coupling dimensions will be referred to in terms of the normalized even-mode impedance. Design systems for various components will be directed toward this key value.

　　Width and gap dimensions may be calculated by de-normalizing the normalized even-mode impedance, calculating the odd-mode impedance and making use of the Cohn nomographs, Figure 4–3 and 4–4 for w/b and s/b.

These nomographs are exact and have been the staple for this calculation for many years, their accuracy limited only by the ability of the user to make proper evaluations from them. Approximate equations[2] based on the fringing capacities in the zero thickness constraint are shown in equations 4–16 and 4–17.

$$Z_{oe} = \frac{94.15 / \sqrt{\epsilon}}{\frac{w}{b} + \frac{\ln 2}{\pi} + \frac{1}{\pi} \ln \left[1 + \tanh \left(\frac{\pi}{2} \cdot \frac{s}{b} \right) \right]} \tag{4-16}$$

$$Z_{oo} = \frac{94.15 / \sqrt{\epsilon}}{\frac{w}{b} + \frac{\ln 2}{\pi} + \frac{1}{\pi} \left[1 + \coth \left(\frac{\pi}{2} \cdot \frac{s}{b} \right) \right]} \tag{4-17}$$

$$\frac{1}{\pi} \ln \left[\tanh \left(\frac{\pi}{2} \cdot \frac{s}{b} \right) \right] = \frac{-188.3 (C_0)}{\sqrt{\epsilon} \quad Z_0 \sqrt{1 - C_0^2}}$$

These, when evaluated as in Equation 4–18, result in

$$\frac{s}{b} = \frac{1}{\pi} \ln \left[\coth \left\{ \frac{94.15 \pi C_0}{\sqrt{\epsilon} \quad Z_0 \sqrt{1 - C_0^2}} \right\} \right] \tag{4-18}$$

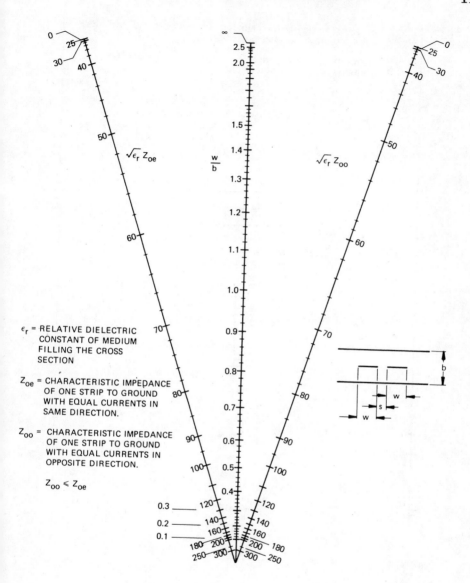

ϵ_r = RELATIVE DIELECTRIC
CONSTANT OF MEDIUM
FILLING THE CROSS
SECTION

Z_{oe} = CHARACTERISTIC IMPEDANCE
OF ONE STRIP TO GROUND
WITH EQUAL CURRENTS IN
SAME DIRECTION.

Z_{oo} = CHARACTERISTIC IMPEDANCE
OF ONE STRIP TO GROUND
WITH EQUAL CURRENTS IN
OPPOSITE DIRECTION.

$Z_{oo} \leqslant Z_{oe}$

FIG. 4–3 Nomogram Giving W/B as a Function of Z_{oe} and Z_{oo} in Coupled Stripline

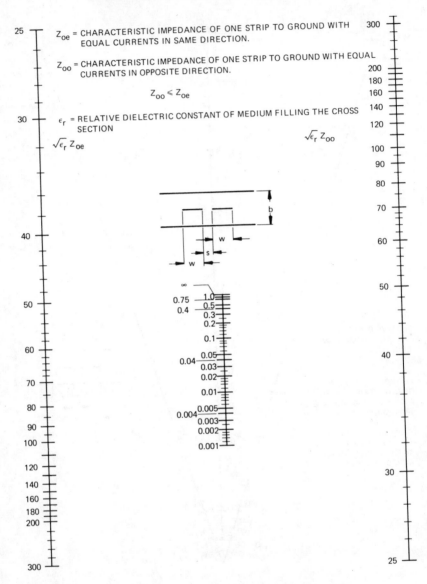

FIG. 4–4   Nomogram giving S/B as a Function of Z_{oe} and Z_{oo} in Coupled Stripline

$$\frac{w}{b} = \frac{94.15}{\sqrt{\epsilon}\ Z_0}\ \sqrt{\frac{1 - C_0}{1 + C_0}}\ - \frac{1}{\pi}\ \log_e \left[2\ \left(1 + e^\theta\right)\right] \tag{4-20}$$

$$\theta = \frac{-188.3\ \pi\ C_0}{\sqrt{\epsilon}\ Z_0\ \sqrt{1 - C_0^2}} \tag{4-21}$$

$$\frac{w}{b} \geq 0.35 \tag{4-22}$$

These are workable relationships which are based only on known constants and on the voltage coupling coefficient, which can be readily calculated from the normalized even-mode impedances as shown in equation 4–13. These design relations have been evaluated for the materials previously described in the Chapter 1, and are represented in Figures 4–5 through 4–9 for values of normalized even-mode impedance from 1 to 2.

Broadside-Coupled Parallel Sections

In much the same way as the side-coupled quarter-wave parallel line sections shown in Figure 4–2a can be described in terms of the even- and odd-mode impedances, so too can the broadside-coupled sections shown in Figure 4–2b. The major differences between these two forms of construction are the degree and overall dynamic range of coupling available for each. The side-coupled structure is primarily for loose coupling and is rarely used for coupling sections tighter than 10 dB, or perhaps, in an extreme case, 6 to 8 dB. On the other hand, the broadside-coupled parallel line structure is useful principally for 3.0 dB or less coupling and is not useful at coupling values much weaker than 6.0 to 8.0 dB. The broadside-coupled structure is, by far, the widest used configuration for 3.0 dB hybrids in practice. Naturally, three-layer, rather than two-layer, construction must be used to achieve this configuration. As before, the basic work was done by Cohn[3], who showed that under the conditions of equation 4–22, the even-mode impedance can be described by:

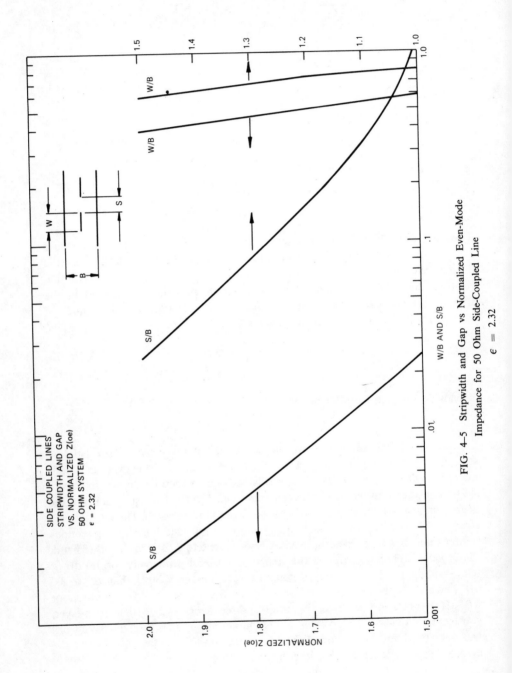

FIG. 4–5 Stripwidth and Gap vs Normalized Even-Mode
Impedance for 50 Ohm Side-Coupled Line
$\epsilon = 2.32$

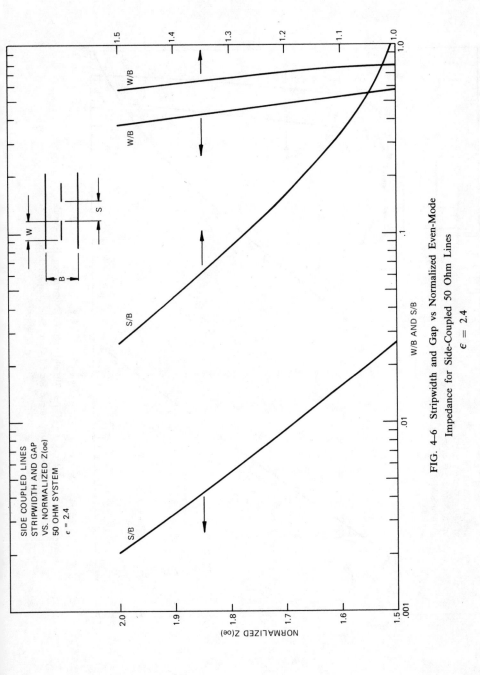

FIG. 4-6 Stripwidth and Gap vs Normalized Even-Mode
Impedance for Side-Coupled 50 Ohm Lines

$\epsilon = 2.4$

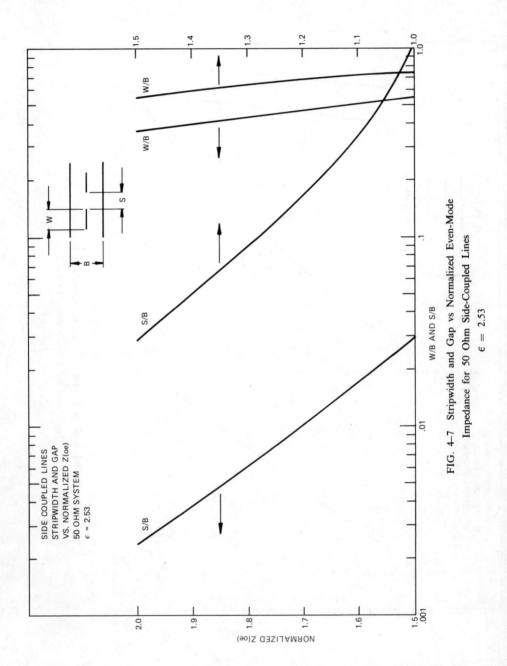

FIG. 4-7 Stripwidth and Gap vs Normalized Even-Mode
Impedance for 50 Ohm Side-Coupled Lines

$\epsilon = 2.53$

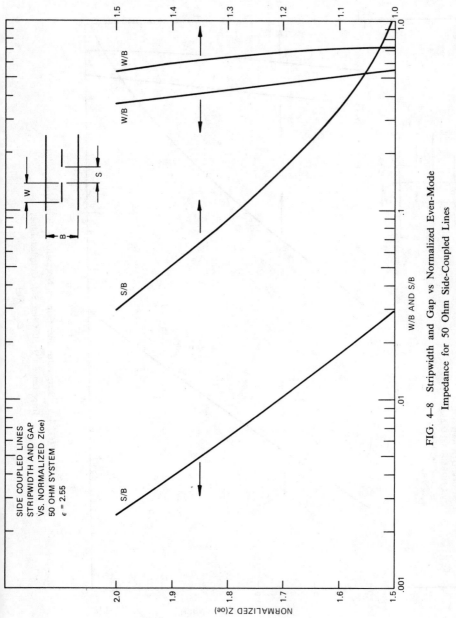

FIG. 4-8 Stripwidth and Gap vs Normalized Even-Mode Impedance for 50 Ohm Side-Coupled Lines

$\epsilon = 2.55$

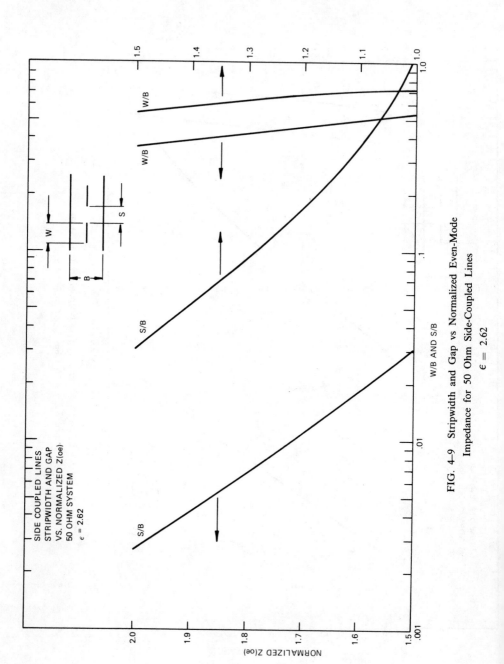

FIG. 4-9 Stripwidth and Gap vs Normalized Even-Mode
Impedance for 50 Ohm Side-Coupled Lines
$\epsilon = 2.62$

$$Z_{oe} = \frac{188.3 / \sqrt{\epsilon}}{\dfrac{w/b}{1 - s/b} + \dfrac{C_{fe}'}{\epsilon}} \tag{4–23}$$

The odd-mode impedance can be described by

$$Z_{oo} = \frac{188.3 / \sqrt{\epsilon}}{\dfrac{w/b}{1 - s/b} + \dfrac{w}{s} + \dfrac{C_{fo}'}{\epsilon}} \tag{4-24}$$

Further work[4] has shown the relationship between the even-mode fringe capacities and the odd-mode fringe capacities for a given gap spacing which is shown in equation 4–25, thus permitting the gap to be dimensioned as a function of the even- and odd-mode impedances by equation 4–26 which leads to the resulting strip width calculation in equation 4–27 where the even-mode fringe capacity is given in 4–28.

$$\frac{C_{fe}' - s/b \quad C_{fo}'}{\epsilon} = 0.4413 \tag{4–25}$$

$$\frac{s}{b} = \frac{Z_{oo}}{Z_{oe}} - \frac{Z_{oo} \sqrt{\epsilon}}{188.3} \left[0.4413 \right] \tag{4–26}$$

$$\frac{w}{b} = \left[\frac{188.3}{Z_{oe} \sqrt{\epsilon}} - \frac{C_{fe}'}{\epsilon} \right] (1 - s/b) \tag{4–27}$$

$$\frac{C_{fe}{}'}{\epsilon} = 0.4413 + \frac{1}{\pi} \left[\log_e \left(\frac{1}{1 - s/b} \right) + \left(\frac{s/b}{1 - s/b} \right) \log_e \frac{b}{s} \right] \qquad (4\text{--}28)$$

These values have been computed for the previously mentioned standard dielectrics and are given in Figures 4–10 through 4–14.

Variable Overlap Coupled Lines

Both the side-coupled structure, previously discussed, and the broadside-coupled structure have severe drawbacks when used in everyday circuit configurations. The side-coupled structure is capable of having only loose coupling characteristics. The broad-coupled structure is at its best only in tight coupling situations. True, the broadside-coupled case can be used for looser values of coupling if the center gap spacing is increased. This, however, is frequently inconvenient when it is combined with other circuit components which require a thin center section shim construction. Additionally, for the intermediate case of coupling in the 5 to 10 dB region, neither of these structures is ideally suited, inasmuch as in one case the gap gets too small to be conveniently fabricated, while in the other case the center-section shim gets too thick to be practical.

Interleaving of the line was suggested by Getsinger[5], and has the advantage of a capability for extremely tight or variable coupling extending to looser coupling values. Its drawback is, however, that it requires a four- or five-layer construction to achieve, and therefore adds substantially to the complexity of any package built using the technique. A practical solution was introduced in 1966 by Shelton[6], who developed a three-layer construction using variable overlap coupled parallel lines as illustrated in Figure 4–15. In this case, it can be seen that for the loose coupling region, the lines do not fully overlap. In a slightly tighter coupling region, the lines may be adjacent or slightly overlapping and for an extremely tight coupling region, the lines are heavily overlapped, reaching the maximum in a full overlap condition. Significantly, this case can be predicted by the same equations as developed by Cohn[3]. This structure, then, solves the problem of a single basic design technique which covers the full range of coupling

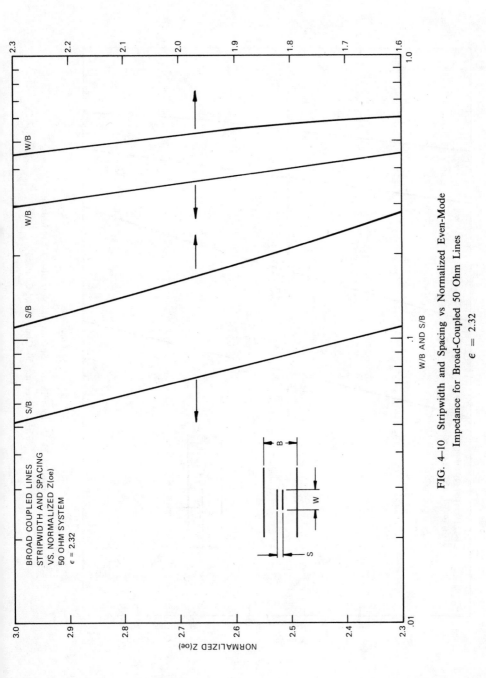

FIG. 4-10 Stripwidth and Spacing vs Normalized Even-Mode
Impedance for Broad-Coupled 50 Ohm Lines

$\epsilon = 2.32$

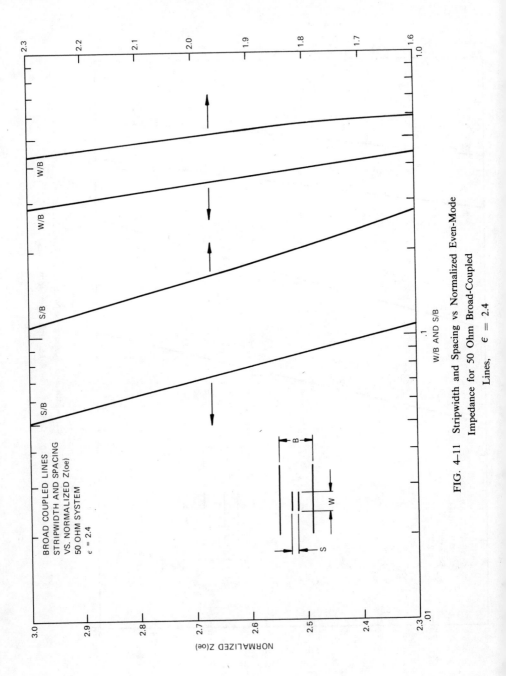

FIG. 4-11 Stripwidth and Spacing vs Normalized Even-Mode
Impedance for 50 Ohm Broad-Coupled
Lines, $\epsilon = 2.4$

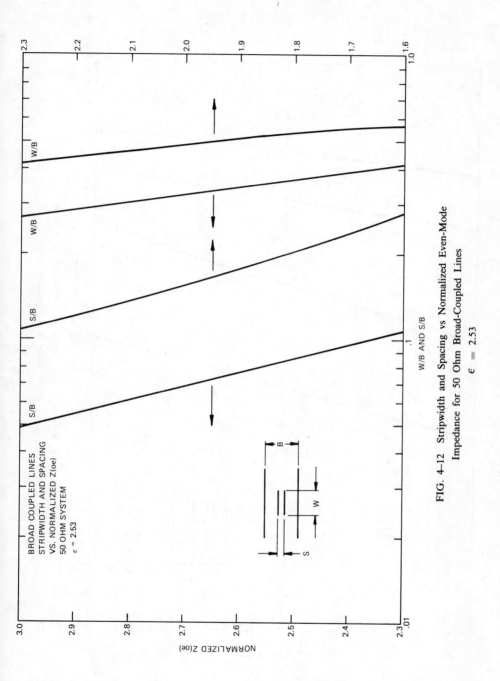

FIG. 4-12 Stripwidth and Spacing vs Normalized Even-Mode
Impedance for 50 Ohm Broad-Coupled Lines
$\epsilon = 2.53$

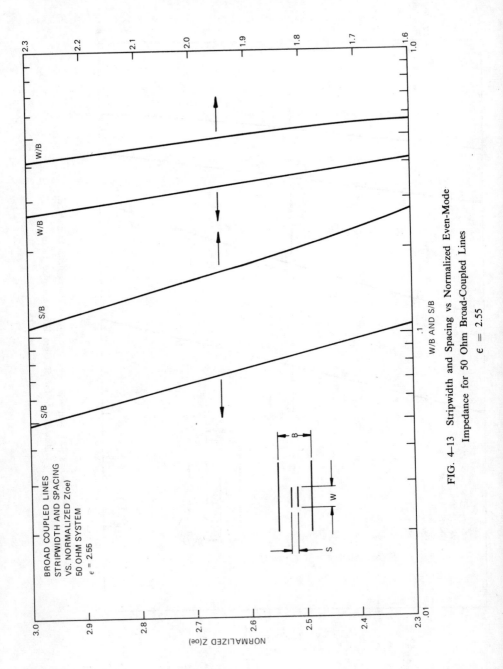

FIG. 4–13 Stripwidth and Spacing vs Normalized Even-Mode
Impedance for 50 Ohm Broad-Coupled Lines
$\epsilon = 2.55$

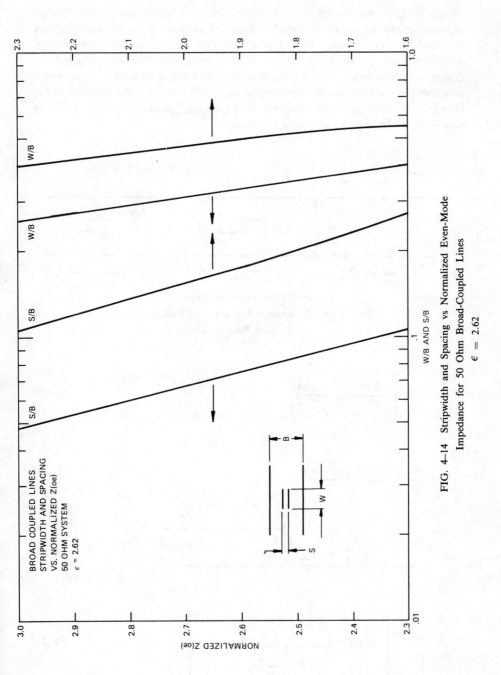

FIG. 4-14 Stripwidth and Spacing vs Normalized Even-Mode
Impedance for 50 Ohm Broad-Coupled Lines

$\epsilon = 2.62$

values within the constraints of construction. Unfortunately, the design dimensions are not nearly as simply calculated as they are for the side-coupled or broadside-coupled cases. In fact, it is difficult to achieve an adequate solution without the use of a digital computer. Shelton recognized this and arranged the design equations in an orderly fashion so that they could be used in a computer program. For completeness, these equations are given in 4–29 through 4–49. However, a thorough understanding of the original principles of the paper is necessary to develop an adequate program.

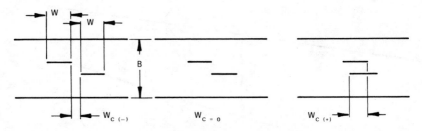

VARIABLE OVERLAP COUPLED PARALLEL LINES

FIG. 4–15 Definition of Parameters of Variable-Overlap
Coupled Parallel Lines

$$A = \exp\left[\frac{60\,\pi^2}{\sqrt{\epsilon}\,Z_o}\left(\frac{1-\rho s}{\sqrt{\rho}}\right)\right] \tag{4-29}$$

$$B = \frac{Pr}{S} = \frac{A - 2 + \sqrt{A^2 - 4A}}{2} \tag{4-30}$$

$$P = \frac{(B-1)\left(\frac{1+s}{2}\right) + \sqrt{\left(\frac{1+s}{2}\right)^2 (B-1)^2 + 4\,sB}}{2} \tag{4-31}$$

$$r = \frac{SB}{P}\quad\left[also, r = \frac{P + \frac{1+s}{2}}{1 + P\left(\frac{1+s}{2S}\right)},\ 0 < P < \infty\right] \tag{4-32}$$

$$C_{fo} = \frac{1}{\pi} \left\{ - \frac{2}{1-S} \log S \right.$$

$$\left. + \frac{1}{S} \log \left[\frac{Pr}{(P+S)\ (1+P)\ (r-S)\ (l-r)} \right] \right\} \qquad (4\text{-}33)$$

$$C_o = \frac{120\ \pi\ \sqrt{\rho}}{\sqrt{\epsilon}\ Z_o} \qquad (4\text{-}34)$$

$$w = \frac{S\ (1-s)}{2} \cdot (C_o - C_{fo}) \qquad (4\text{-}35)$$

$$w_c = \frac{1}{2\ \pi} \left[(1+S)\ \log \frac{P}{r} + (l-S)\ \log \left[\left(\frac{l+P}{S+P} \right) \left(\frac{r-s}{l-r} \right) \right] \right] \qquad (4\text{-}36)$$

$$\frac{1 - \rho_{max.}}{\sqrt{\rho_{max}}}\ S = \frac{\sqrt{\epsilon}\ Z_o}{60\ \pi^2}\ \log 4 \qquad (4\text{-}37)$$

$$C_o = \frac{120\ \pi\ \sqrt{\rho}}{\sqrt{\epsilon}\ Z_o} \qquad (4\text{-}38)$$

$$\triangle C = \frac{120\ \pi}{\sqrt{\epsilon}\ Z_o}\ \frac{\rho - 1}{\sqrt{\rho}} \qquad (4\text{-}39)$$

$$k = \frac{1}{\exp \dfrac{\pi \triangle C}{2} - 1} \qquad (4\text{-}40)$$

$$a = + \sqrt{\left(\frac{s-k}{s+1}\right)^2 + k - \left(\frac{s-k}{s+1}\right)} \qquad (4\text{-}41)$$

$$q = \frac{k}{a} \left[\text{ or } q = \left(\frac{s+1}{2}\right) \left\{ \frac{a + \dfrac{2s}{s+1}}{a + \dfrac{s+1}{2}} \right\} \right] \qquad (4\text{-}42)$$

$$C_{fo} = \frac{2}{\pi} \left[\frac{1}{1+s} \log\left(\frac{1+a}{a(1-q)}\right) - \frac{1}{1-s} \log q \right] \qquad (4\text{-}43)$$

$$w_c = \frac{1}{\pi} \left[s \log \frac{q}{a} + (1\text{-}s) \log\left(\frac{1-q}{1-a}\right) \right] \qquad (4\text{-}44)$$

$$C_{f(a=\infty)} = -\frac{2}{\pi} \left[\frac{1}{1+s} \log\left(\frac{1-s}{2}\right) + \frac{1}{1-s} \log\left(\frac{1+s}{2}\right) \right] \qquad (4\text{-}45)$$

$$w = \frac{1-s^2}{4} \left[C_o - C_{fo} - C_{f(a=\infty)} \right] \qquad (4\text{-}46)$$

$$\frac{w}{1-S} \geq 0.35 \qquad (4\text{-}47)$$

$$\frac{2W_o}{1+s} \geq 0.85 \qquad (4\text{-}48)$$

$$\frac{w_c}{s} \geq 0.7 \qquad\qquad\qquad (4\text{--}49)$$

Because of the extremely wide dynamic range of the normalized even-mode impedances which can be achieved with this structure, it is difficult to plot the design parameters in a graph, as has been done with the other types of construction. A partially successful attempt was done[7] in a larger format for some of the basic materials and aspect ratios of shim to total ground-plane spacing, but the curves are extremely difficult to read to the necessary accuracy needed for some of the more sophisticated multi-section devices which are now in common usage. The problem exists to an even greater extent in attempting to provide curves for this book, and, for this reason, none have been included. Rather, design tables for each of the basic dielectric materials specified, as well as the three different shim thicknesses and ground-plane spacings, are included in this chapter. The number of points in the tables has been increased at both coupling extremes, and are reduced in the central section, which is approximately a straight line function. To use them, plot the area of specific interest on a greatly expanded scale which will then permit interpolation between the computer derived points. Referring again to Figure 4–15, it should be noted that a negative value of $w_{(c)}$ means that there is a gap between the lines while a positive value means that there is an overlap between the lines. The limiting case of this overlap occurs where the line width and the overlap are equal. This data is contained in Tables 4–1 through 4–5.

This chapter has presented the three major coupling structures in use today for parallel quarter-wave coupled line sections. In each case, design data and equations have been presented which relate this to the normalized even-mode impedance of the section. Throughout the remainder of this book, designs will be brought to the level of the normalized even-mode impedance where appropriate and may then be referred to this chapter for translation into physical dimensions making use of whichever of the three coupling structures is most appropriate.

DIELECTRIC CONSTANT = 2.32
SHIM THICKNESS = .005
BASE MATERIAL = .031
TOTAL G.P.S. = .067

NORM. Z(OE)	WIDTH	OVERLAP
1.0100	.0530	-.0692
1.0200	.0530	-.0546
1.0300	.0529	-.0461
1.0400	.0528	-.0401
1.0500	.0527	-.0355
1.0600	.0526	-.0318
1.0700	.0525	-.0287
1.0800	.0523	-.0261
1.0900	.0522	-.0238
1.1000	.0520	-.0218
1.2000	.0495	-.0100
1.3000	.0464	-.0045
1.4000	.0432	-.0014
1.5000	.0403	-.0009
1.6000	.0377	.0028
1.7000	.0355	.0047
1.8000	.0336	.0064
1.9000	.0319	.0081
2.0000	.0305	.0098
2.1000	.0292	.0115
2.2000	.0281	.0132
2.3000	.0272	.0149
2.4000	.0264	.0167
2.5000	.0257	.0187
2.6000	.0251	.0213
2.6100	.0250	.0216
2.6200	.0249	.0220
2.6300	.0249	.0224
2.6400	.0248	.0229
2.6500	.0248	.0236

DIELECTRIC CONSTANT = 2.32
SHIM THICKNESS = .005
BASE MATERIAL = .062
TOTAL G.P.S. = .129

NORM. Z(OE)	WIDTH	OVERLAP
1.0100	.1025	-.1335
1.0200	.1024	-.1053
1.0300	.1023	-.0889
1.0400	.1022	-.0774
1.0500	.1020	-.0686
1.0600	.1018	-.0615
1.0700	.1015	-.0556
1.0800	.1012	-.0506
1.0900	.1009	-.0462
1.1000	.1005	-.0424
1.2000	.0956	-.0199
1.3000	.0894	-.0100
1.4000	.0829	-.0049
1.5000	.0768	-.0017
1.6000	.0712	-.0005
1.7000	.0662	.0023
1.8000	.0619	.0042
1.9000	.0580	.0059
2.0000	.0545	.0075
2.1000	.0515	.0091
2.2000	.0487	.0107
2.3000	.0463	.0122
2.4000	.0441	.0138
2.5000	.0422	.0154
2.6000	.0404	.0170
2.7000	.0388	.0186
2.8000	.0374	.0202
2.9000	.0361	.0219
3.0000	.0349	.0237
3.1000	.0338	.0256
3.2000	.0329	.0280
3.2100	.0328	.0283
3.2200	.0327	.0286
3.2300	.0326	.0289
3.2400	.0326	.0293
3.2500	.0325	.0297
3.2600	.0324	.0303
3.2700	.0323	.0310

DIELECTRIC CONSTANT = 2.32
SHIM THICKNESS = .005
BASE MATERIAL = .125
TOTAL G.P.S. = .255

NORM. Z(OE)	WIDTH	OVERLAP
1.0100	.2029	-.2640
1.0200	.2028	-.2082
1.0300	.2026	-.1759
1.0400	.2023	-.1532
1.0500	.2019	-.1358
1.0600	.2014	-.1218
1.0700	.2009	-.1101
1.0800	.2003	-.1002
1.0900	.1996	-.0916
1.1000	.1989	-.0840
1.2000	.1892	-.0397
1.3000	.1768	-.0205
1.4000	.1638	-.0109
1.5000	.1512	-.0056
1.6000	.1396	-.0024
1.7000	.1291	-.0001
1.8000	.1197	.0018
1.9000	.1114	.0037
2.0000	.1038	.0053
2.1000	.0971	.0068
2.2000	.0910	.0084
2.3000	.0855	.0099
2.4000	.0805	.0115
2.5000	.0759	.0130
2.6000	.0718	.0145
2.7000	.0680	.0160
2.8000	.0645	.0175
2.9000	.0613	.0190
3.0000	.0583	.0205
3.1000	.0556	.0220
3.2000	.0531	.0236
3.3000	.0508	.0251
3.4000	.0486	.0267
3.5000	.0466	.0283
3.6000	.0448	.0301
3.7000	.0431	.0319
3.8000	.0415	.0340
3.9000	.0400	.0373
3.9100	.0399	.0380

DIELECTRIC CONSTANT = 2.32
SHIM THICKNESS = .011
BASE MATERIAL = .031
TOTAL G.P.S. = .073

NORM. Z(OE)	WIDTH	OVERLAP
1.0100	.0566	-.0750
1.0200	.0566	-.0590
1.0300	.0565	-.0497
1.0400	.0564	-.0432
1.0500	.0563	-.0381
1.0600	.0562	-.0341
1.0700	.0561	-.0306
1.0800	.0559	-.0277
1.0900	.0557	-.0252
1.1000	.0555	-.0230
1.2000	.0530	-.0093
1.3000	.0501	-.0021
1.4000	.0473	-.0029
1.5000	.0450	.0077
1.6000	.0429	.0118
1.7000	.0413	.0160
1.8000	.0399	.0202
1.9000	.0388	.0248
2.0000	.0380	.0305
2.0100	.0379	.0312
2.0200	.0378	.0321
2.0300	.0377	.0330
2.0400	.0377	.0343
2.0500	.0376	.0364

DIELECTRIC CONSTANT = 2.32
SHIM THICKNESS = .011
BASE MATERIAL = .062
TOTAL G.P.S. = .135

NORM. Z(OE)	WIDTH	OVERLAP
1.0100	.1066	-.1395
1.0200	.1066	-.1099
1.0300	.1065	-.0928
1.0400	.1063	-.0808
1.0500	.1061	-.0715
1.0600	.1059	-.0641
1.0700	.1056	-.0578
1.0800	.1053	-.0525
1.0900	.1050	-.0479
1.1000	.1046	-.0439
1.2000	.0996	-.0199
1.3000	.0934	-.0088
1.4000	.0871	-.0022
1.5000	.0814	-.0026
1.6000	.0764	-.0067
1.7000	.0720	.0110
1.8000	.0683	.0148
1.9000	.0650	.0185
2.0000	.0622	.0222
2.1000	.0598	.0260
2.2000	.0578	.0298
2.3000	.0560	.0338
2.4000	.0545	.0381
2.5000	.0531	.0433
2.5100	.0530	.0440
2.5200	.0529	.0447
2.5300	.0528	.0454
2.5400	.0527	.0462
2.5500	.0526	.0471
2.5600	.0525	.0482
2.5700	.0524	.0498

DIELECTRIC CONSTANT = 2.32
SHIM THICKNESS = .011
BASE MATERIAL = .125
TOTAL G.P.S. = .261

NORM. Z(OE)	WIDTH	OVERLAP
1.0100	.2074	-.2701
1.0200	.2072	-.2130
1.0300	.2070	-.1799
1.0400	.2067	-.1567
1.0500	.2063	-.1388
1.0600	.2058	-.1245
1.0700	.2053	-.1125
1.0800	.2047	-.1023
1.0900	.2040	-.0935
1.1000	.2033	-.0857
1.2000	.1934	-.0401
1.3000	.1808	-.0201
1.4000	.1678	-.0095
1.5000	.1555	-.0030
1.6000	.1443	-.0017
1.7000	.1344	.0058
1.8000	.1257	.0099
1.9000	.1179	.0135
2.0000	.1111	.0171
2.1000	.1050	.0206
2.2000	.0996	.0241
2.3000	.0948	.0276
2.4000	.0905	.0311
2.5000	.0867	.0346
2.6000	.0832	.0381
2.7000	.0801	.0417
2.8000	.0773	.0454
2.9000	.0748	.0492
3.0000	.0726	.0534
3.1000	.0705	.0584
3.1100	.0704	.0589
3.1200	.0702	.0596
3.1300	.0700	.0602
3.1400	.0698	.0609
3.1500	.0696	.0616
3.1600	.0694	.0625
3.1700	.0693	.0634
3.1800	.0691	.0646
3.1900	.0689	.0663

DIELECTRIC CONSTANT = 2.32
SHIM THICKNESS = .022
BASE MATERIAL = .031
TOTAL G.P.S. = .084

NORM. Z(OE)	WIDTH	OVERLAP
1.0100	.0616	-.0850
1.0200	.0616	-.0664
1.0300	.0615	-.0556
1.0400	.0614	-.0480
1.0500	.0613	-.0421
1.0600	.0612	-.0372
1.0700	.0611	-.0332
1.0800	.0609	-.0297
1.0900	.0607	-.0266
1.1000	.0606	-.0239
1.2000	.0583	-.0060
1.3000	.0559	.0051
1.4000	.0539	.0161
1.5000	.0522	.0259
1.6000	.0511	.0379
1.6100	.0510	.0395
1.6200	.0509	.0414
1.6300	.0508	.0437
1.6400	.0507	.0473

DIELECTRIC CONSTANT = 2.32
SHIM THICKNESS = .022
BASE MATERIAL = .062
TOTAL G.P.S. = .146

NORM. Z(OE)	WIDTH	OVERLAP
1.0100	.1132	-.1500
1.0200	.1131	-.1180
1.0300	.1130	-.0994
1.0400	.1128	-.0863
1.0500	.1126	-.0762
1.0600	.1124	-.0681
1.0700	.1121	-.0613
1.0800	.1118	-.0555
1.0900	.1114	-.0504
1.1000	.1111	-.0459
1.2000	.1061	-.0185
1.3000	.1002	-.0042
1.4000	.0947	.0058
1.5000	.0899	.0155
1.6000	.0858	.0237
1.7000	.0825	.0319
1.8000	.0798	.0404
1.9000	.0776	.0495
2.0000	.0759	.0609
2.0100	.0758	.0625
2.0200	.0756	.0641
2.0300	.0755	.0661
2.0400	.0753	.0685
2.0500	.0752	.0727

DIELECTRIC CONSTANT = 2.32
SHIM THICKNESS = .022
BASE MATERIAL = .125
TOTAL G.P.S. = .272

NORM. Z(OE)	WIDTH	OVERLAP
1.0100	.2149	-.2810
1.0200	.2148	-.2215
1.0300	.2145	-.1870
1.0400	.2142	-.1687
1.0500	.2138	-.1441
1.0600	.2133	-.1291
1.0700	.2128	-.1166
1.0800	.2122	-.1059
1.0900	.2115	-.0966
1.1000	.2107	-.0885
1.2000	.2007	-.0401
1.3000	.1882	-.0178
1.4000	.1755	-.0046
1.5000	.1640	.0051
1.6000	.1538	.0133
1.7000	.1450	.0219
1.8000	.1374	.0294
1.9000	.1309	.0369
2.0000	.1253	.0443
2.1000	.1204	.0519
2.2000	.1162	.0595
2.3000	.1126	.0674
2.4000	.1095	.0760
2.5000	.1068	.0862
2.5100	.1066	.0875
2.5200	.1063	.0887
2.5300	.1061	.0901
2.5400	.1059	.0916
2.5500	.1057	.0933
2.5600	.1054	.0952
2.5700	.1052	.0977
2.5800	.1050	.1015

DIELECTRIC CONSTANT = 2.4
SHIM THICKNESS = .005
BASE MATERIAL = .062
TOTAL G.P.S. = .129

NORM. Z(OE)	WIDTH	OVERLAP
1.0100	.0998	-.1342
1.0200	.0998	-.1060
1.0300	.0997	-.0896
1.0400	.0995	-.0781
1.0500	.0993	-.0693
1.0600	.0991	-.0622
1.0700	.0989	-.0563
1.0800	.0986	-.0513
1.0900	.0983	-.0469
1.1000	.0979	-.0430
1.2000	.0932	-.0204
1.3000	.0872	-.0104
1.4000	.0809	-.0052
1.5000	.0749	-.0020
1.6000	.0694	-.0002
1.7000	.0646	.0021
1.8000	.0603	.0040
1.9000	.0564	.0056
2.0000	.0530	.0072
2.1000	.0500	.0088
2.2000	.0474	.0103
2.3000	.0450	.0119
2.4000	.0428	.0134
2.5000	.0409	.0150
2.6000	.0391	.0166
2.7000	.0376	.0181
2.8000	.0362	.0198
2.9000	.0349	.0214
3.0000	.0337	.0232
3.1000	.0327	.0252
3.2000	.0318	.0278
3.2100	.0317	.0281
3.2200	.0316	.0285
3.2300	.0315	.0290
3.2400	.0314	.0295
3.2500	.0313	.0305

DIELECTRIC CONSTANT = 2.4
SHIM THICKNESS = .005
BASE MATERIAL = .031
TOTAL G.P.S. = .067

NORM. Z(OE)	WIDTH	OVERLAP
1.0100	.0516	-.0696
1.0200	.0516	-.0549
1.0300	.0515	-.0464
1.0400	.0514	-.0405
1.0500	.0514	-.0359
1.0600	.0512	-.0322
1.0700	.0511	-.0291
1.0800	.0510	-.0265
1.0900	.0508	-.0242
1.1000	.0506	-.0222
1.2000	.0482	-.0102
1.3000	.0452	-.0048
1.4000	.0421	-.0016
1.5000	.0393	.0007
1.6000	.0368	.0025
1.7000	.0346	.0045
1.8000	.0327	.0062
1.9000	.0311	.0078
2.0000	.0296	.0095
2.1000	.0284	.0111
2.2000	.0273	.0128
2.3000	.0264	.0146
2.4000	.0256	.0164
2.5000	.0249	.0184
2.6000	.0243	.0211
2.6100	.0243	.0215
2.6200	.0242	.0219
2.6300	.0241	.0225
2.6400	.0241	.0234
3.8100	.0396	.0340
3.8200	.0394	.0343
3.8300	.0393	.0347
3.8400	.0391	.0350
3.8500	.0390	.0354
3.8600	.0388	.0359
3.8700	.0387	.0365
3.8800	.0385	.0374

DIELECTRIC CONSTANT = 2.4
SHIM THICKNESS = .005
BASE MATERIAL = .125
TOTAL G.P.S. = .255

NORM. Z(OE)	WIDTH	OVERLAP
1.0100	.1976	-.2654
1.0200	.1975	-.2096
1.0300	.1973	-.1773
1.0400	.1970	-.1546
1.0500	.1966	-.1372
1.0600	.1962	-.1231
1.0700	.1957	-.1114
1.0800	.1951	-.1015
1.0900	.1945	-.0928
1.1000	.1938	-.0853
1.2000	.1844	-.0407
1.3000	.1724	-.0212
1.4000	.1597	-.0114
1.5000	.1475	-.0060
1.6000	.1362	-.0027
1.7000	.1259	-.0004
1.8000	.1167	.0015
1.9000	.1084	.0032
2.0000	.1010	.0050
2.1000	.0944	.0065
2.2000	.0884	.0080
2.3000	.0830	.0096
2.4000	.0781	.0111
2.5000	.0736	.0126
2.6000	.0695	.0140
2.7000	.0658	.0155
2.8000	.0623	.0170
2.9000	.0592	.0185
3.0000	.0563	.0200
3.1000	.0536	.0215
3.2000	.0512	.0230
3.3000	.0489	.0246
3.4000	.0468	.0262
3.5000	.0448	.0278
3.6000	.0430	.0295
3.7000	.0413	.0314
3.8000	.0397	.0338

Table 1

DIELECTRIC CONSTANT = 2.4
SHIM THICKNESS = .011
BASE MATERIAL = .031
TOTAL G.P.S. = .073

NORM. Z(OE)	WIDTH	OVERLAP
1.0100	.0551	-.0754
1.0200	.0551	-.0594
1.0300	.0550	-.0501
1.0400	.0549	-.0435
1.0500	.0548	-.0385
1.0600	.0547	-.0344
1.0700	.0546	-.0310
1.0800	.0544	-.0281
1.0900	.0543	-.0256
1.1000	.0541	-.0233
1.2000	.0517	-.0096
1.3000	.0488	-.0024
1.4000	.0461	.0026
1.5000	.0438	.0068
1.6000	.0418	.0114
1.7000	.0402	.0155
1.8000	.0388	.0197
1.9000	.0378	.0243
2.0000	.0369	.0301
2.0100	.0368	.0309
2.0200	.0368	.0319
2.0300	.0367	.0330
2.0400	.0366	.0347

Table 2

DIELECTRIC CONSTANT = 2.4
SHIM THICKNESS = .011
BASE MATERIAL = .062
TOTAL G.P.S. = .135

NORM. Z(OE)	WIDTH	OVERLAP
1.0100	.1039	-.1402
1.0200	.1038	-.1107
1.0300	.1037	-.0935
1.0400	.1035	-.0815
1.0500	.1033	-.0722
1.0600	.1031	-.0648
1.0700	.1029	-.0585
1.0800	.1026	-.0532
1.0900	.1022	-.0486
1.1000	.1019	-.0446
1.2000	.0970	-.0205
1.3000	.0910	-.0093
1.4000	.0850	-.0026
1.5000	.0793	.0022
1.6000	.0744	.0062
1.7000	.0702	.0105
1.8000	.0665	.0142
1.9000	.0633	.0179
2.0000	.0605	.0216
2.1000	.0582	.0253
2.2000	.0561	.0291
2.3000	.0544	.0331
2.4000	.0529	.0375
2.5000	.0516	.0428
2.5100	.0515	.0435
2.5200	.0514	.0443
2.5300	.0512	.0451
2.5400	.0511	.0460
2.5500	.0510	.0472
2.5600	.0509	.0491

Table 3

DIELECTRIC CONSTANT = 2.4
SHIM THICKNESS = .011
BASE MATERIAL = .125
TOTAL G.P.S. = .261

NORM. Z(OE)	WIDTH	OVERLAP
1.0100	.2019	-.2715
1.0200	.2018	-.2144
1.0300	.2016	-.1813
1.0400	.2013	-.1581
1.0500	.2009	-.1402
1.0600	.2005	-.1258
1.0700	.2000	-.1139
1.0800	.1994	-.1037
1.0900	.1987	-.0948
1.1000	.1980	-.0870
1.2000	.1885	-.0412
1.3000	.1763	-.0209
1.4000	.1637	-.0101
1.5000	.1516	-.0035
1.6000	.1407	.0012
1.7000	.1310	.0052
1.8000	.1224	.0094
1.9000	.1148	.0129
2.0000	.1081	.0164
2.1000	.1021	.0199
2.2000	.0968	.0234
2.3000	.0921	.0268
2.4000	.0878	.0302
2.5000	.0841	.0337
2.6000	.0807	.0372
2.7000	.0776	.0408
2.8000	.0749	.0444
2.9000	.0724	.0483
3.0000	.0702	.0526
3.1000	.0682	.0578
3.1100	.0680	.0584
3.1200	.0678	.0591
3.1300	.0676	.0598
3.1400	.0675	.0607
3.1500	.0673	.0617
3.1600	.0671	.0629
3.1700	.0669	.0648

DIELECTRIC CONSTANT = 2.4
SHIM THICKNESS = .022
BASE MATERIAL = .031
TOTAL G.P.S. = .084

NORM. Z(OE)	WIDTH	OVERLAP
1.0100	.0600	-.0854
1.0200	.0599	-.0669
1.0300	.0599	-.0561
1.0400	.0598	-.0484
1.0500	.0597	-.0425
1.0600	.0596	-.0377
1.0700	.0595	-.0336
1.0800	.0593	-.0301
1.0900	.0591	-.0271
1.1000	.0590	-.0243
1.2000	.0568	-.0064
1.3000	.0544	-.0046
1.4000	.0525	.0156
1.5000	.0509	.0253
1.6000	.0497	.0375
1.6100	.0496	.0392
1.6200	.0495	.0412
1.6300	.0494	.0439

DIELECTRIC CONSTANT = 2.4
SHIM THICKNESS = .022
BASE MATERIAL = .062
TOTAL G.P.S. = .146

NORM. Z(OE)	WIDTH	OVERLAP
1.0100	.1102	-.1508
1.0200	.1102	-.1188
1.0300	.1100	-.1002
1.0400	.1099	-.0871
1.0500	.1097	-.0770
1.0600	.1095	-.0689
1.0700	.1092	-.0621
1.0800	.1089	-.0562
1.0900	.1085	-.0512
1.1000	.1082	-.0467
1.2000	.1034	-.0192
1.3000	.0977	-.0049
1.4000	.0923	.0051
1.5000	.0875	.0136
1.6000	.0836	.0229
1.7000	.0803	.0310
1.8000	.0777	.0394
1.9000	.0755	.0486
2.0000	.0738	.0603
2.0100	.0737	.0619
2.0200	.0736	.0637
2.0300	.0734	.0660
2.0400	.0733	.0694

DIELECTRIC CONSTANT = 2.4
SHIM THICKNESS = .022
BASE MATERIAL = .125
TOTAL G.P.S. = .272

NORM. Z(OE)	WIDTH	OVERLAP
1.0100	.2093	-.2825
1.0200	.2091	-.2230
1.0300	.2089	-.1884
1.0400	.2086	-.1642
1.0500	.2082	-.1455
1.0600	.2078	-.1305
1.0700	.2073	-.1180
1.0800	.2067	-.1073
1.0900	.2060	-.0980
1.1000	.2053	-.0898
1.2000	.1956	-.0412
1.3000	.1834	-.0187
1.4000	.1712	-.0054
1.5000	.1598	.0042
1.6000	.1499	.0124
1.7000	.1413	.0209
1.8000	.1338	.0284
1.9000	.1274	.0357
2.0000	.1218	.0431
2.1000	.1170	.0505
2.2000	.1129	.0581
2.3000	.1094	.0660
2.4000	.1063	.0746
2.5000	.1037	.0851
2.5100	.1035	.0865
2.5200	.1032	.0879
2.5300	.1030	.0894
2.5400	.1028	.0911
2.5500	.1025	.0932
2.5600	.1023	.0958

DIELECTRIC CONSTANT = 2.53
SHIM THICKNESS = 0.005
BASE MATERIAL = 0.031
TOTAL G.P.S. = 0.067

NORM. Z(OE)	WIDTH	OVERLAP
1.0100	.0495	-.0702
1.0200	.0495	-.0555
1.0300	.0494	-.0470
1.0400	.0493	-.0410
1.0500	.0493	-.0364
1.0600	.0492	-.0327
1.0700	.0490	-.0296
1.0800	.0489	-.0270
1.0900	.0487	-.0247
1.1000	.0486	-.0227
1.2000	.0463	-.0107
1.3000	.0434	-.0051
1.4000	.0405	-.0019
1.5000	.0377	-.0004
1.6000	.0353	.0022
1.7000	.0332	.0041
1.8000	.0313	.0058
1.9000	.0297	.0074
2.0000	.0284	.0090
2.1000	.0272	.0107
2.2000	.0261	.0123
2.3000	.0252	.0141
2.4000	.0244	.0159
2.5000	.0238	.0180
2.6000	.0232	.0210
2.6100	.0231	.0216

DIELECTRIC CONSTANT = 2.53
SHIM THICKNESS = 0.005
BASE MATERIAL = 0.062
TOTAL G.P.S. = 0.129

NORM. Z(OE)	WIDTH	OVERLAP
1.0100	.0958	-.1353
1.0200	.0957	-.1071
1.0300	.0956	-.0907
1.0400	.0955	-.0792
1.0500	.0953	-.0704
1.0600	.0951	-.0633
1.0700	.0948	-.0573
1.0800	.0946	-.0523
1.0900	.0943	-.0479
1.1000	.0939	-.0440
1.2000	.0895	-.0212
1.3000	.0838	-.0110
1.4000	.0778	-.0056
1.5000	.0720	-.0024
1.6000	.0667	-.0001
1.7000	.0620	.0017
1.8000	.0578	.0036
1.9000	.0541	.0052
2.0000	.0508	.0067
2.1000	.0478	.0083
2.2000	.0452	.0098
2.3000	.0429	.0113
2.4000	.0408	.0129
2.5000	.0389	.0144
2.6000	.0372	.0159
2.7000	.0357	.0175
2.8000	.0343	.0191
2.9000	.0331	.0208
3.0000	.0320	.0226
3.1000	.0310	.0247
3.2000	.0301	.0279
3.2100	.0300	.0286

DIELECTRIC CONSTANT = 2.53
SHIM THICKNESS = 0.005
BASE MATERIAL = 0.125
TOTAL G.P.S. = 0.255

NORM. Z(OE)	WIDTH	OVERLAP
1.0100	.1895	-.2675
1.0200	.1894	-.2117
1.0300	.1892	-.1794
1.0400	.1890	-.1567
1.0500	.1886	-.1393
1.0600	.1882	-.1252
1.0700	.1877	-.1135
1.0800	.1872	-.1035
1.0900	.1866	-.0948
1.1000	.1859	-.0872
1.2000	.1770	-.0423
1.3000	.1656	-.0224
1.4000	.1536	-.0123
1.5000	.1418	-.0066
1.6000	.1309	-.0032
1.7000	.1209	-.0008
1.8000	.1120	.0011
1.9000	.1039	.0027
2.0000	.0968	.0045
2.1000	.0903	.0060
2.2000	.0845	.0075
2.3000	.0792	.0090
2.4000	.0744	.0105
2.5000	.0700	.0119
2.6000	.0661	.0134
2.7000	.0624	.0148
2.8000	.0591	.0163
2.9000	.0560	.0178
3.0000	.0532	.0192
3.1000	.0506	.0207
3.2000	.0482	.0222
3.3000	.0460	.0238
3.4000	.0439	.0253
3.5000	.0420	.0270
3.6000	.0402	.0288
3.7000	.0386	.0308
3.8000	.0370	.0338
3.8100	.0369	.0344
3.8200	.0368	.0351

DIELECTRIC CONSTANT = 2.53
SHIM THICKNESS = 0.011
BASE MATERIAL = 0.031
TOTAL G.P.S. = 0.073

NORM. Z(OE)	WIDTH	OVERLAP
1.0100	.0529	-.0760
1.0200	.0528	-.0600
1.0300	.0528	-.0507
1.0400	.0527	-.0442
1.0500	.0526	-.0391
1.0600	.0525	-.0350
1.0700	.0524	-.0316
1.0800	.0522	-.0287
1.0900	.0521	-.0262
1.1000	.0519	-.0239
1.2000	.0496	-.0101
1.3000	.0469	-.0029
1.4000	.0443	-.0020
1.5000	.0420	.0062
1.6000	.0401	.0108
1.7000	.0385	.0148
1.8000	.0372	.0190
1.9000	.0362	.0236
2.0000	.0354	.0297
2.0100	.0353	.0307
2.0200	.0352	.0319
2.0300	.0351	.0340

DIELECTRIC CONSTANT = 2.53
SHIM THICKNESS = 0.011
BASE MATERIAL = 0.062
TOTAL G.P.S. = 0.135

NORM. Z(OE)	WIDTH	OVERLAP
1.0100	.0996	-.1413
1.0200	.0996	-.1118
1.0300	.0994	-.0946
1.0400	.0993	-.0826
1.0500	.0991	-.0733
1.0600	.0989	-.0659
1.0700	.0987	-.0596
1.0800	.0984	-.0543
1.0900	.0981	-.0497
1.1000	.0977	-.0456
1.2000	.0932	-.0213
1.3000	.0875	-.0100
1.4000	.0816	-.0033
1.5000	.0762	.0015
1.6000	.0714	.0055
1.7000	.0673	.0098
1.8000	.0637	.0134
1.9000	.0606	.0170
2.0000	.0579	.0207
2.1000	.0556	.0243
2.2000	.0536	.0281
2.3000	.0519	.0320
2.4000	.0505	.0364
2.5000	.0492	.0422
2.5100	.0491	.0431
2.5200	.0490	.0440
2.5300	.0489	.0452
2.5400	.0488	.0472

DIELECTRIC CONSTANT = 2.53
SHIM THICKNESS = 0.011
BASE MATERIAL = 0.125
TOTAL G.P.S. = 0.261

NORM. Z(OE)	WIDTH	OVERLAP
1.0100	.1937	-.2737
1.0200	.1936	-.2166
1.0300	.1934	-.1835
1.0400	.1931	-.1602
1.0500	.1927	-.1424
1.0600	.1923	-.1280
1.0700	.1918	-.1160
1.0800	.1913	-.1057
1.0900	.1907	-.0969
1.1000	.1900	-.0890
1.2000	.1810	-.0428
1.3000	.1694	-.0221
1.4000	.1573	-.0111
1.5000	.1458	-.0044
1.6000	.1352	-.0004
1.7000	.1257	-.0044
1.8000	.1174	-.0085
1.9000	.1100	-.0120
2.0000	.1034	-.0155
2.1000	.0976	-.0189
2.2000	.0925	-.0222
2.3000	.0879	-.0256
2.4000	.0837	-.0290
2.5000	.0800	-.0324
2.6000	.0767	-.0358
2.7000	.0738	-.0394
2.8000	.0711	-.0430
2.9000	.0687	-.0469
3.0000	.0665	-.0513
3.1000	.0646	-.0573
3.1100	.0644	-.0581
3.1200	.0642	-.0592
3.1300	.0641	-.0606

DIELECTRIC CONSTANT = 2.53
SHIM THICKNESS = 0.022
BASE MATERIAL = 0.031
TOTAL G.P.S. = 0.084

NORM. Z(OE)	WIDTH	OVERLAP
1.0100	.0575	-.0861
1.0200	.0575	-.0676
1.0300	.0574	-.0568
1.0400	.0573	-.0492
1.0500	.0572	-.0432
1.0600	.0571	-.0384
1.0700	.0570	-.0344
1.0800	.0569	-.0309
1.0900	.0567	-.0278
1.1000	.0565	-.0250
1.2000	.0545	-.0072
1.3000	.0522	.0037
1.4000	.0503	.0125
1.5000	.0488	.0244
1.6000	.0477	.0369
1.6100	.0476	.0388
1.6200	.0475	.0413
1.6300	.0474	.0459

DIELECTRIC CONSTANT = 2.53
SHIM THICKNESS = 0.022
BASE MATERIAL = 0.062
TOTAL G.P.S. = 0.146

NORM. Z(OE)	WIDTH	OVERLAP
1.0100	.1057	-.1520
1.0200	.1056	-.1200
1.0300	.1055	-.1014
1.0400	.1054	-.0883
1.0500	.1052	-.0782
1.0600	.1050	-.0701
1.0700	.1047	-.0633
1.0800	.1044	-.0574
1.0900	.1041	-.0523
1.1000	.1038	-.0478
1.2000	.0992	-.0202
1.3000	.0938	-.0059
1.4000	.0886	.0041
1.5000	.0840	.0124
1.6000	.0802	.0216
1.7000	.0770	.0297
1.8000	.0744	.0380
1.9000	.0724	.0471
2.0000	.0707	.0594
2.0100	.0706	.0613
2.0200	.0704	.0638
2.0300	.0703	.0680

DIELECTRIC CONSTANT = 2.53
SHIM THICKNESS = 0.022
BASE MATERIAL = 0.125
TOTAL G.P.S. = 0.272

NORM. Z(OE)	WIDTH	OVERLAP
1.0100	.2007	-.2848
1.0200	.2006	-.2252
1.0300	.2004	-.1907
1.0400	.2001	-.1664
1.0500	.1997	-.1478
1.0600	.1993	-.1327
1.0700	.1988	-.1202
1.0800	.1983	-.1095
1.0900	.1976	-.1002
1.1000	.1970	-.0920
1.2000	.1878	-.0430
1.3000	.1762	-.0202
1.4000	.1645	-.0067
1.5000	.1535	.0029
1.6000	.1439	.0110
1.7000	.1355	.0194
1.8000	.1282	.0267
1.9000	.1220	.0340
2.0000	.1166	.0412
2.1000	.1119	.0485
2.2000	.1079	.0560
2.3000	.1044	.0639
2.4000	.1015	.0726
2.5000	.0989	.0838
2.5100	.0987	.0854
2.5200	.0985	.0871
2.5300	.0982	.0891
2.5400	.0980	.0918

DIELECTRIC CONSTANT = 2.55
SHIM THICKNESS = 0.005
BASE MATERIAL = 0.031
TOTAL G.P.S. = 0.067

NORM. Z(OE)	WIDTH	OVERLAP
1.0100	.0492	-.0703
1.0200	.0492	-.0556
1.0300	.0491	-.0471
1.0400	.0490	-.0411
1.0500	.0490	-.0365
1.0600	.0488	-.0328
1.0700	.0487	-.0297
1.0800	.0486	-.0271
1.0900	.0484	-.0248
1.1000	.0483	-.0228
1.2000	.0460	-.0107
1.3000	.0432	-.0052
1.4000	.0402	-.0019
1.5000	.0375	-.0003
1.6000	.0351	.0022
1.7000	.0330	.0041
1.8000	.0311	.0057
1.9000	.0296	.0074
2.0000	.0282	.0090
2.1000	.0270	.0106
2.2000	.0259	.0123
2.3000	.0250	.0140
2.4000	.0243	.0158
2.5000	.0236	.0179
2.6000	.0230	.0211
2.6100	.0230	.0217

DIELECTRIC CONSTANT = 2.55
SHIM THICKNESS = 0.005
BASE MATERIAL = 0.062
TOTAL G.P.S. = 0.129

NORM. Z(OE)	WIDTH	OVERLAP
1.0100	.0952	-.1354
1.0200	.0951	-.1072
1.0300	.0950	-.0909
1.0400	.0949	-.0794
1.0500	.0947	-.0705
1.0600	.0945	-.0634
1.0700	.0943	-.0575
1.0800	.0940	-.0524
1.0900	.0937	-.0480
1.1000	.0934	-.0442
1.2000	.0889	-.0213
1.3000	.0833	-.0111
1.4000	.0773	-.0057
1.5000	.0716	-.0025
1.6000	.0663	-.0002
1.7000	.0616	.0017
1.8000	.0574	.0035
1.9000	.0537	.0051
2.0000	.0504	.0067
2.1000	.0475	.0082
2.2000	.0449	.0097
2.3000	.0426	.0113
2.4000	.0405	.0128
2.5000	.0386	.0143
2.6000	.0370	.0158
2.7000	.0354	.0174
2.8000	.0341	.0190
2.9000	.0328	.0207
3.0000	.0317	.0225
3.1000	.0307	.0246
3.2000	.0298	.0280
3.2100	.0297	.0291

DIELECTRIC CONSTANT = 2.55
SHIM THICKNESS = 0.005
BASE MATERIAL = 0.125
TOTAL G.P.S. = 0.255

NORM. Z(OE)	WIDTH	OVERLAP
1.0100	.1884	-.2678
1.0200	.1882	-.2121
1.0300	.1880	-.1797
1.0400	.1878	-.1570
1.0500	.1874	-.1396
1.0600	.1870	-.1255
1.0700	.1866	-.1138
1.0800	.1860	-.1038
1.0900	.1854	-.0951
1.1000	.1848	-.0875
1.2000	.1760	-.0425
1.3000	.1646	-.0226
1.4000	.1526	-.0124
1.5000	.1410	-.0067
1.6000	.1301	-.0033
1.7000	.1202	-.0009
1.8000	.1113	.0010
1.9000	.1033	.0026
2.0000	.0961	.0044
2.1000	.0897	.0059
2.2000	.0839	.0074
2.3000	.0786	.0089
2.4000	.0739	.0104
2.5000	.0695	.0118
2.6000	.0656	.0133
2.7000	.0619	.0147
2.8000	.0586	.0162
2.9000	.0555	.0177
3.0000	.0527	.0191
3.1000	.0501	.0206
3.2000	.0477	.0221
3.3000	.0455	.0236
3.4000	.0435	.0252
3.5000	.0416	.0269
3.6000	.0398	.0287
3.7000	.0382	.0308
3.8000	.0367	.0340
3.8100	.0365	.0347

DIELECTRIC CONSTANT = 2.55
SHIM THICKNESS = 0.011
BASE MATERIAL = 0.031
TOTAL G.P.S. = 0.073

NORM. Z(OE)	WIDTH	OVERLAP
1.0100	.0525	-.0761
1.0200	.0525	-.0601
1.0300	.0524	-.0508
1.0400	.0524	-.0442
1.0500	.0523	-.0392
1.0600	.0522	-.0351
1.0700	.0520	-.0317
1.0800	.0519	-.0288
1.0900	.0517	-.0263
1.1000	.0516	-.0240
1.2000	.0493	-.0102
1.3000	.0466	-.0030
1.4000	.0440	-.0020
1.5000	.0417	.0061
1.6000	.0399	.0107
1.7000	.0383	.0147
1.8000	.0370	.0189
1.9000	.0359	.0235
2.0000	.0351	.0297
2.0100	.0350	.0307
2.0200	.0350	.0319

DIELECTRIC CONSTANT = 2.55
SHIM THICKNESS = 0.011
BASE MATERIAL = 0.062
TOTAL G.P.S. = 0.135

NORM. Z(OE)	WIDTH	OVERLAP
1.0100	.0990	-.1415
1.0200	.0989	-.1120
1.0300	.0988	-.0948
1.0400	.0987	-.0828
1.0500	.0985	-.0735
1.0600	.0983	-.0660
1.0700	.0981	-.0598
1.0800	.0978	-.0545
1.0900	.0975	-.0499
1.1000	.0971	-.0458
1.2000	.0926	-.0215
1.3000	.0869	-.0101
1.4000	.0811	-.0034
1.5000	.0758	-.0014
1.6000	.0710	.0054
1.7000	.0669	.0097
1.8000	.0633	.0133
1.9000	.0602	.0169
2.0000	.0575	.0205
2.1000	.0553	.0242
2.2000	.0533	.0279
2.3000	.0516	.0319
2.4000	.0501	.0363
2.5000	.0489	.0422
2.5100	.0488	.0430
2.5200	.0486	.0440
2.5300	.0485	.0454

DIELECTRIC CONSTANT = 2.55
SHIM THICKNESS = 0.011
BASE MATERIAL = 0.125
TOTAL G.P.S. = 0.261

NORM. Z(OE)	WIDTH	OVERLAP
1.0100	.1925	-.2740
1.0200	.1924	-.2169
1.0300	.1922	-.1838
1.0400	.1919	-.1605
1.0500	.1915	-.1427
1.0600	.1911	-.1283
1.0700	.1906	-.1163
1.0800	.1901	-.1060
1.0900	.1895	-.0972
1.1000	.1888	-.0894
1.2000	.1799	-.0431
1.3000	.1684	-.0223
1.4000	.1564	-.0113
1.5000	.1449	-.0045
1.6000	.1344	-.0003
1.7000	.1249	.0043
1.8000	.1167	.0084
1.9000	.1093	.0119
2.0000	.1028	.0153
2.1000	.0970	.0187
2.2000	.0918	.0221
2.3000	.0872	.0254
2.4000	.0831	.0288
2.5000	.0795	.0322
2.6000	.0762	.0356
2.7000	.0732	.0392
2.8000	.0705	.0428
2.9000	.0682	.0467
3.0000	.0660	.0511
3.1000	.0641	.0573
3.1100	.0639	.0582
3.1200	.0637	.0594
3.1300	.0635	.0611

DIELECTRIC CONSTANT = 2.55
SHIM THICKNESS = 0.022
BASE MATERIAL = 0.031
TOTAL G.P.S. = 0.084

NORM. Z(OE)	WIDTH	OVERLAP
1.0100	.0571	-.0862
1.0200	.0571	-.0677
1.0300	.0570	-.0569
1.0400	.0570	-.0493
1.0500	.0569	-.0433
1.0600	.0568	-.0385
1.0700	.0566	-.0345
1.0800	.0565	-.0310
1.0900	.0564	-.0279
1.1000	.0562	-.0252
1.2000	.0541	-.0073
1.3000	.0519	-.0036
1.4000	.0500	.0124
1.5000	.0485	.0242
1.6000	.0474	.0368
1.6100	.0473	.0388
1.6200	.0472	.0413

DIELECTRIC CONSTANT = 2.55
SHIM THICKNESS = 0.022
BASE MATERIAL = 0.062
TOTAL G.P.S. = 0.146

NORM. Z(OE)	WIDTH	OVERLAP
1.0100	.1050	-.1522
1.0200	.1050	-.1202
1.0300	.1049	-.1016
1.0400	.1047	-.0885
1.0500	.1045	-.0784
1.0600	.1043	-.0703
1.0700	.1041	-.0634
1.0800	.1038	-.0576
1.0900	.1035	-.0525
1.1000	.1031	-.0480
1.2000	.0986	-.0204
1.3000	.0932	-.0060
1.4000	.0880	.0039
1.5000	.0835	.0122
1.6000	.0797	.0215
1.7000	.0765	.0295
1.8000	.0740	.0378
1.9000	.0719	.0469
2.0000	.0702	.0593
2.0100	.0701	.0613
2.0200	.0700	.0639

DIELECTRIC CONSTANT = 2.55
SHIM THICKNESS = 0.022
BASE MATERIAL = 0.125
TOTAL G.P.S. = 0.272

NORM. Z(OE)	WIDTH	OVERLAP
1.0100	.1995	-.2851
1.0200	.1993	-.2256
1.0300	.1991	-.1910
1.0400	.1989	-.1668
1.0500	.1985	-.1481
1.0600	.1981	-.1331
1.0700	.1976	-.1205
1.0800	.1970	-.1098
1.0900	.1964	-.1005
1.1000	.1957	-.0923
1.2000	.1866	-.0433
1.3000	.1752	-.0204
1.4000	.1635	-.0069
1.5000	.1526	-.0027
1.6000	.1430	.0108
1.7000	.1347	.0192
1.8000	.1274	.0265
1.9000	.1212	.0337
2.0000	.1158	.0409
2.1000	.1112	.0482
2.2000	.1072	.0557
2.3000	.1037	.0636
2.4000	.1008	.0723
2.5000	.0982	.0837
2.5100	.0980	.0853
2.5200	.0978	.0871
2.5300	.0975	.0893
2.5400	.0973	.0924

DIELECTRIC CONSTANT = 2.62
SHIM THICKNESS = 0.005
BASE MATERIAL = 0.062
TOTAL G.P.S. = 0.129

NORM. Z(OE)	WIDTH	OVERLAP
1.0100	.0931	-.1360
1.0200	.0931	-.1078
1.0300	.0930	-.0914
1.0400	.0928	-.0799
1.0500	.0927	-.0711
1.0600	.0925	-.0640
1.0700	.0922	-.0580
1.0800	.0920	-.0530
1.0900	.0917	-.0486
1.1000	.0914	-.0447
1.2000	.0871	-.0217
1.3000	.0815	-.0114
1.4000	.0757	-.0060
1.5000	.0701	-.0027
1.6000	.0649	-.0004
1.7000	.0603	.0015
1.8000	.0562	.0031
1.9000	.0525	.0049
2.0000	.0493	.0065
2.1000	.0464	.0080
2.2000	.0439	.0095
2.3000	.0416	.0110
2.4000	.0395	.0125
2.5000	.0377	.0140
2.6000	.0360	.0155
2.7000	.0345	.0171
2.8000	.0332	.0187
2.9000	.0319	.0204
3.0000	.0308	.0222
3.1000	.0298	.0244
3.1100	.0297	.0247
3.1200	.0297	.0249
3.1300	.0296	.0252
3.1400	.0295	.0255
3.1500	.0294	.0259
3.1600	.0293	.0263
3.1700	.0292	.0267
3.1800	.0291	.0273
3.1900	.0290	.0282

DIELECTRIC CONSTANT = 2.62
SHIM THICKNESS = 0.005
BASE MATERIAL = 0.031
TOTAL G.P.S. = 0.067

NORM. Z(OE)	WIDTH	OVERLAP
1.0100	.0481	-.0705
1.0200	.0481	-.0559
1.0300	.0481	-.0474
1.0400	.0480	-.0414
1.0500	.0479	-.0368
1.0600	.0478	-.0331
1.0700	.0477	-.0300
1.0800	.0475	-.0274
1.0900	.0474	-.0251
1.1000	.0472	-.0230
1.2000	.0450	-.0110
1.3000	.0423	-.0053
1.4000	.0394	-.0021
1.5000	.0367	.0002
1.6000	.0344	.0020
1.7000	.0323	.0039
1.8000	.0305	.0056
1.9000	.0289	.0072
2.0000	.0275	.0088
2.1000	.0264	.0104
2.2000	.0253	.0120
2.3000	.0244	.0137
2.4000	.0237	.0156
2.5000	.0230	.0177
2.6000	.0224	.0214
3.7300	.0364	.0314
3.7400	.0362	.0318
3.7500	.0360	.0321
3.7600	.0359	.0325
3.7700	.0358	.0330
3.7800	.0356	.0337
3.7900	.0355	.0349

DIELECTRIC CONSTANT = 2.62
SHIM THICKNESS = 0.005
BASE MATERIAL = 0.125
TOTAL G.P.S. = 0.255

NORM. Z(OE)	WIDTH	OVERLAP
1.0100	.1843	-.2689
1.0200	.1842	-.2132
1.0300	.1840	-.1808
1.0400	.1837	-.1581
1.0500	.1834	-.1406
1.0600	.1830	-.1266
1.0700	.1826	-.1149
1.0800	.1820	-.1048
1.0900	.1815	-.0962
1.1000	.1808	-.0885
1.2000	.1723	-.0433
1.3000	.1613	-.0232
1.4000	.1495	-.0129
1.5000	.1381	-.0071
1.6000	.1274	-.0035
1.7000	.1177	-.0011
1.8000	.1089	-.0008
1.9000	.1010	.0024
2.0000	.0940	.0042
2.1000	.0876	.0057
2.2000	.0819	.0072
2.3000	.0767	.0086
2.4000	.0720	.0101
2.5000	.0677	.0115
2.6000	.0638	.0130
2.7000	.0602	.0144
2.8000	.0570	.0158
2.9000	.0540	.0173
3.0000	.0512	.0188
3.1000	.0486	.0202
3.2000	.0463	.0217
3.3000	.0441	.0233
3.4000	.0420	.0248
3.5000	.0402	.0265
3.6000	.0384	.0284
3.7000	.0368	.0306
3.7100	.0367	.0308
3.7200	.0365	.0311

DIELECTRIC CONSTANT = 2.62
SHIM THICKNESS = 0.011
BASE MATERIAL = 0.031
TOTAL G.P.S. = 0.073

NORM. Z(OE)	WIDTH	OVERLAP
1.0100	.0514	-.0764
1.0200	.0514	-.0604
1.0300	.0513	-.0511
1.0400	.0512	-.0446
1.0500	.0511	-.0395
1.0600	.0510	-.0354
1.0700	.0509	-.0320
1.0800	.0508	-.0291
1.0900	.0506	-.0266
1.1000	.0505	-.0243
1.2000	.0483	-.0105
1.3000	.0456	-.0033
1.4000	.0431	-.0017
1.5000	.0409	.0058
1.6000	.0390	.0104
1.7000	.0374	.0144
1.8000	.0362	.0185
1.9000	.0351	.0231
2.0000	.0343	.0296
2.0100	.0343	.0307
2.0200	.0342	.0325

DIELECTRIC CONSTANT = 2.62
SHIM THICKNESS = 0.011
BASE MATERIAL = 0.062
TOTAL G.P.S. = 0.135

NORM. Z(OE)	WIDTH	OVERLAP
1.0100	.0969	-.1421
1.0200	.0968	-.1125
1.0300	.0967	-.0954
1.0400	.0966	-.0833
1.0500	.0964	-.0741
1.0600	.0962	-.0666
1.0700	.0960	-.0604
1.0800	.0957	-.0550
1.0900	.0954	-.0504
1.1000	.0951	-.0463
1.2000	.0907	-.0219
1.3000	.0851	-.0105
1.4000	.0795	-.0037
1.5000	.0742	.0011
1.6000	.0695	.0051
1.7000	.0655	.0093
1.8000	.0619	.0129
1.9000	.0589	.0165
2.0000	.0562	.0201
2.1000	.0540	.0237
2.2000	.0520	.0274
2.3000	.0504	.0314
2.4000	.0489	.0358
2.5000	.0477	.0421
2.5100	.0476	.0431
2.5200	.0475	.0444

DIELECTRIC CONSTANT = 2.62
SHIM THICKNESS = 0.011
BASE MATERIAL = 0.125
TOTAL G.P.S. = 0.261

NORM. Z(OE)	WIDTH	OVERLAP
1.0100	.1883	-.2751
1.0200	.1882	-.2180
1.0300	.1880	-.1849
1.0400	.1878	-.1616
1.0500	.1874	-.1438
1.0600	.1870	-.1294
1.0700	.1866	-.1174
1.0800	.1860	-.1071
1.0900	.1855	-.0982
1.1000	.1848	-.0904
1.2000	.1761	-.0439
1.3000	.1650	-.0230
1.4000	.1532	-.0118
1.5000	.1419	-.0049
1.6000	.1316	-.0001
1.7000	.1223	.0039
1.8000	.1142	.0080
1.9000	.1069	.0114
2.0000	.1004	.0148
2.1000	.0947	.0182
2.2000	.0897	.0215
2.3000	.0851	.0248
2.4000	.0811	.0282
2.5000	.0775	.0315
2.6000	.0742	.0350
2.7000	.0713	.0385
2.8000	.0687	.0421
2.9000	.0663	.0461
3.0000	.0642	.0506
3.1000	.0623	.0576
3.1100	.0621	.0591

DIELECTRIC CONSTANT = 2.62
SHIM THICKNESS = 0.022
BASE MATERIAL = 0.031
TOTAL G.P.S. = 0.084

NORM. Z(OE)	WIDTH	OVERLAP
1.0100	.0559	-.0866
1.0200	.0559	-.0681
1.0300	.0558	-.0573
1.0400	.0557	-.0496
1.0500	.0556	-.0437
1.0600	.0555	-.0389
1.0700	.0554	-.0348
1.0800	.0553	-.0313
1.0900	.0551	-.0283
1.1000	.0550	-.0255
1.2000	.0530	-.0077
1.3000	.0508	.0032
1.4000	.0489	.0119
1.5000	.0474	.0238
1.6000	.0463	.0366
1.6100	.0462	.0387
1.6200	.0461	.0417

DIELECTRIC CONSTANT = 2.62
SHIM THICKNESS = 0.022
BASE MATERIAL = 0.062
TOTAL G.P.S. = 0.146

NORM. Z(OE)	WIDTH	OVERLAP
1.0100	.1028	-.1529
1.0200	.1027	-.1208
1.0300	.1026	-.1022
1.0400	.1025	-.0891
1.0500	.1023	-.0790
1.0600	.1021	-.0709
1.0700	.1018	-.0641
1.0800	.1016	-.0582
1.0900	.1013	-.0531
1.1000	.1009	-.0486
1.2000	.0966	-.0209
1.3000	.0913	-.0066
1.4000	.0862	.0034
1.5000	.0817	.0116
1.6000	.0780	.0209
1.7000	.0749	.0288
1.8000	.0723	.0370
1.9000	.0703	.0462
2.0000	.0687	.0591
2.0100	.0685	.0614
2.0200	.0684	.0650

DIELECTRIC CONSTANT = 2.62
SHIM THICKNESS = 0.022
BASE MATERIAL = 0.125
TOTAL G.P.S. = 0.272

NORM. Z(OE)	WIDTH	OVERLAP
1.0100	.1952	-.2863
1.0200	.1951	-.2268
1.0300	.1949	-.1922
1.0400	.1946	-.1679
1.0500	.1942	-.1493
1.0600	.1938	-.1342
1.0700	.1934	-.1216
1.0800	.1928	-.1109
1.0900	.1922	-.1016
1.1000	.1916	-.0934
1.2000	.1827	-.0442
1.3000	.1716	-.0212
1.4000	.1601	-.0076
1.5000	.1494	-.0020
1.6000	.1400	.0100
1.7000	.1318	.0185
1.8000	.1246	.0257
1.9000	.1185	.0328
2.0000	.1132	.0400
2.1000	.1086	.0472
2.2000	.1047	.0547
2.3000	.1013	.0625
2.4000	.0983	.0713
2.5000	.0958	.0833
2.5100	.0956	.0851
2.5200	.0954	.0874
2.5300	.0952	.0906

REFERENCES

1 Cohn, Seymour *Shielded Coupled Strip Transmission Line,* MTT-5, October, 1955, pp. 29–37.

2 Richardson, J. K., *Graphical Design of Stripline Directional Couplers,* Microwaves, October 1967, pp. 71–74.

3 Cohn, Seymour, *Characteristic Impedance of Broad Side Coupled Strip Transmission Lines,* MTT-8, No. 6, November 1960, pp. 633–637.

4 Gunderson, L. C. and Guida, A., *Stripline Coupler Design,* Microwave Journal, Vol. 8, No. 6, June 1956, pp. 97–101.

5 Getsinger, W.J., *A Coupled Stripline Configuration Using Printed Circuit Construction That Allows Very Close Coupling,* MTT-9, No. 6, November 1961, pp. 535–544.

6 Shelton, P. J., *Impedances of Offset Parallel Coupled Strip Transmission Line,* MTT-14, No. 1, January 1966, pp. 7–15.

7 Mosco, J. A., *Coupling Curves for Offset Parallel Coupled Strip Transmission Line,* Microwave Journal, Vol. 10, No. 5., April 1967, pp. 35–37.

CHAPTER
—5——————————————————

Parallel Coupled Line Directional Couplers

Broadband directional couplers can be constructed by having a main transmission line in parallel proximity to a secondary line. As a result of this proximity, a percentage of the power present on the main line is coupled to the secondary line; the power varies as a function of the physical dimensions of the coupler and the direction of the propagation of the primary power. This may be seen in Figure 5–1a. Power incident on the main line at port J1 will be coupled at some reduced power level to port J2, while the primary power continues to port J4. Port J3 is normally terminated in a load equal to the characteristic impedance of the line; in an ideal coupler, no power will appear there. The amount of power coupled from port J1 to J2 is a function of the construction of the coupler as expressed by the even- and odd-mode impedances previously discussed in Chapter 4. The relationship of this coupler is shown by equations 5–1 and 5–2.

$$Z_{oe} = Z_o \sqrt{\frac{1 + C_0}{1 - C_0}} \qquad (5\text{--}1)$$

$$Z_{oo} = Z_o \sqrt{\frac{1 - C_0}{1 + C_0}} \qquad (5\text{--}2)$$

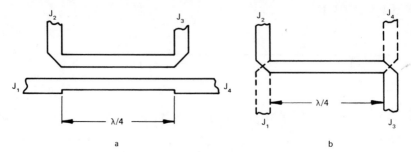

FIG. 5–1 Quarter-Wave Coupled-Line Directional Coupler
Configurations

This is related to the voltage coupling coefficient, K, as defined in equation 5–3 at frequencies at which the coupled length is an odd multiple of λ/4. As previously mentioned, for a matched condition, equation 5–4 must be satisfied.

$$\text{Coupling (dB)} = 20 \, \text{Log}_{10} \, C_0 \qquad\qquad (5\text{–}3)$$

$$Z_o = \sqrt{Z_{oe} \, Z_{oo}} \qquad\qquad (5\text{–}4)$$

Therefore, if Z_o is normalized to one, it can be seen that the coupling can be expressed in terms of only the normalized even-mode impedance, for which design parameters have already been presented.

Figure 5–2 is a plot of normalized even-mode impedance versus coupling expressed in dB.

The usable bandwidth of a single-section, quarter-wave coupler is approximately one octave. The typical responses of several values of coupling versus normalized frequency are shown in Figure 5–3.

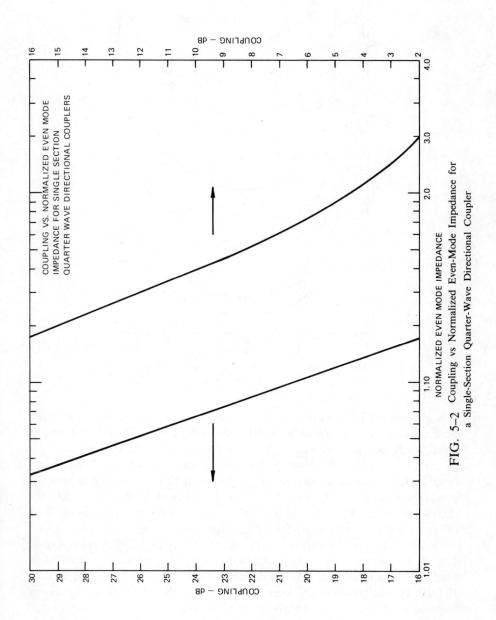

FIG. 5-2 Coupling vs Normalized Even-Mode Impedance for a Single-Section Quarter-Wave Directional Coupler

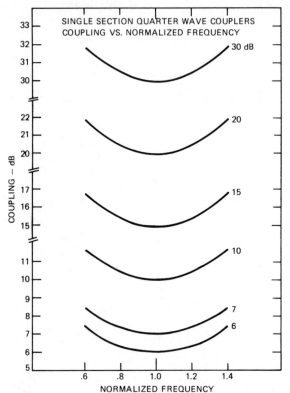

FIG. 5–3 Frequency Response for a Number of Single-Section
Quarter-Wave Couplers

It is obvious from these curves that the coupling is weaker at both ends of the band. In order to maximize the effective usable bandwidth of such a device, it is frequently desirable to overcouple at the design frequency, thus permitting a plus and minus tolerance across the frequency range, varying about the nominal coupling value. The exact amount of coupling shift varies according to both the nominal coupling and the desired bandwidth and is difficult to read from the curves presented. Table 5–1 is a chart of center-coupling values, with their variation in dB for several selected bandwidths and a variety of coupling values. Sufficient data is given to permit interpolation between the points. As an example, a 10 dB directional coupler built as a single-section device for use over a 40% bandwidth should be designed for a nominal coupling value of 9.8 dB rather than 10.0 dB. This will result in a $\pm$ tolerance of 0.2 dB, rather than a center frequency response of 10.0 dB and a 0.4 dB error at the band edges.

DESIGN VALUE SHIFT FOR SINGLE SECTION QUARTER WAVE COUPLERS

NOMINAL COUPLING	10% BW		20% BW		30% BW		40% BW		66% BW	
	DESIGN COUPLING	±dB	DESIGN COUPLING	±dB	DESIGN COUPLING	±dB	DESIGN COUPLING	±dB	DESIGN COUPLING	±dB
3	2.99	.007	2.97	.026	2.94	.06	2.89	.11	2.68	.32
6	5.99	.01	5.96	.040	5.91	.09	5.83	.16	5.52	.48
7	6.99	.011	6.96	.043	6.90	.1	6.82	.17	6.49	.51
10	9.99	.012	9.95	.048	9.89	.11	9.80	.2	9.42	.57
13	12.99	.013	12.95	.051	12.88	.12	12.79	.21	12.39	.61
15	14.99	.013	14.95	.052	14.88	.12	14.79	.21	14.38	.62
17	16.98	.013	16.95	.053	16.88	.12	16.79	.21	16.37	.63
20	19.98	.013	19.95	.053	19.88	.12	19.78	.22	19.36	.63
23	22.98	.013	22.95	.054	22.88	.12	22.78	.22	22.36	.63
27	26.98	.013	26.95	.054	26.88	.12	26.78	.22	26.36	.64
30	29.98	.013	29.95	.054	29.88	.12	29.78	.22	29.36	.64

TABLE 5-1

Directivity

Directivity is a quality factor related to the directional coupler; it defines the amount of power appearing at the uncoupled port, J3. In absolute terms, it is expressed as isolation; however, this is not a true measure of coupling quality. Directivity is equal to isolation minus coupling as shown in equation 5–5, and therefore remains relatively constant as a function of physical construction rather than of coupling unless the even- and odd-mode velocities are unequal.

Directivity = (Isolation) - (Coupling)   (5–5)

This condition occurs more frequently with weak coupling, but is not normally a problem with tight couplers. There are several limiting and controlling factors on directivity. One factor is the uniformity of propagation of the even- and odd-mode waves within the coupler. The second is the internal match of the coupler which is a function of the balance of these even- and odd-mode impedances. The third factor is the end junction mismatches which result as the secondary line is decoupled from the primary line at each end of the quarter-wave section; and a fourth is the load VSWR on the main and secondary output ports.

Assuming construction in a solid homogeneous medium, without air gaps, VSWR becomes the most frequently limiting parameter on coupler directivity. Its relationship is shown in Figure 5–4, which is actually a plot of return

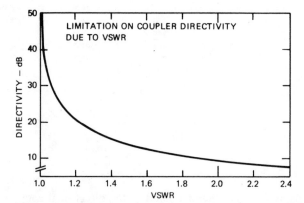

FIG. 5–4 Limitation of Coupler Directivity Due to VSWR

loss, expressed in dB, as described in equation 5–6.

$$\text{Return Loss (dB)} = -20 \, \text{Log}_{10} \left[\frac{\text{VSWR} - 1}{\text{VSWR} + 1} \right] \qquad (5\text{-}6)$$

By control of the load impedances as well as by careful matching of the end discontinuities, it is possible to build stripline directional couplers with directivities in the 30 dB region. For production purposes, most directional couplers are best specified at directivity values in the order of 20 dB. These values will also decrease as the frequency increases since the discontinuities are far more difficult to control at higher frequencies.

3 dB Case

For an equal-power split, a 3.0 dB (3.01 dB) directional coupler may be employed. The single-section quarter-wave directional coupler has a 90° phase relationship between the outputs; this is useful for many applications. For circuit layout reasons which will become evident in later chapters, it is generally laid out as a completely overlapped coupler with a cross-over, so that the arms are as shown in Figure 5–1b.
Input power at J1 splits equally between ports J2 and J4 with a 90° phase relationship between these two outputs; J3 is the isolated arm. As in the case of the looser values of coupling, it is frequently desirable to overcouple at the center frequency in order to provide a $\pm$ tolerance around the nominal design frequency. The theoretical frequency response of several 3.0 dB directional couplers is shown in Figure 5–5 for both the main and coupled arms.

Construction

For coupling values weaker than 10.0 dB, the most convenient method of construction is the side coupled configuration as described in Chapter 4. As coupling values become tighter than 10.0 dB, the gap between the lines becomes very small. It becomes difficult to etch and control, resulting in a design which is not practical for production use. Thus, for values less than 10.0 dB, it is frequently desirable to use the partial offset technique.

Multi-Section Couplers

For increased bandwidth and flatness, the single-section quarter-wave directional coupler may be expanded to any number of sections, limited only by practicality and the desire to make use of symmetric or asymmetric construction. Symmetric couplers will always have an odd number of sections and will be, as the name implies, symmetrical around the center section, which is always the

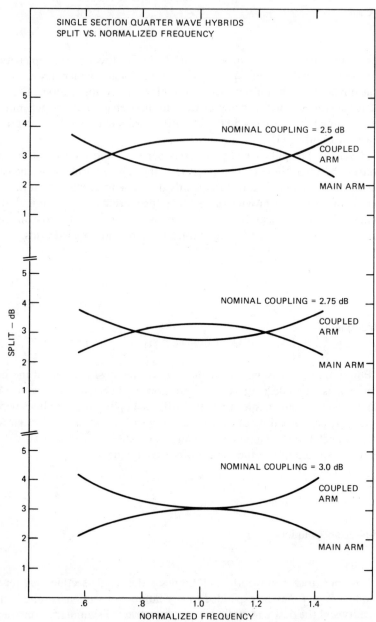

FIG. 5–5 Frequency Response for Several Single-Section
Quarter-Wave 3.0 dB Hybrids for Varying
Values of Nominal Coupling

most tightly coupled section. A representation of the symmetric multi-section coupler is shown in Figure 5–6.

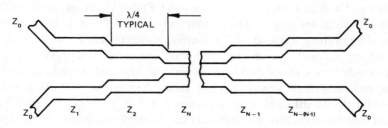

FIG. 5–6   Symmetric Multi-Section Directional Coupler

The frequency response of this type of device is an equal-ripple function as shown in Figure 5–7. Here, the response of several 10 dB couplers designed for a ripple of $\pm$ 0.2 dB have been plotted versus both frequency and the number of sections used.

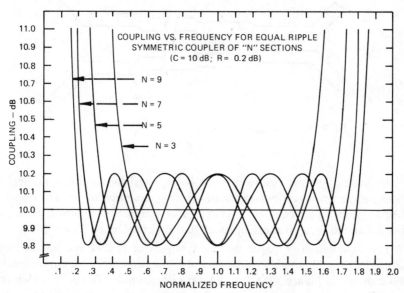

FIG. 5–7   Coupling vs Frequency for Equal-Ripple Symmetric
Directional Couplers Having Three to Nine Sections

As would be expected, the bandwidth increases with the number of sections and also with the amount of ripple.

Papers describing this type of coupler have been prepared by a variety of authors, and may be found in the bibliography. Perhaps the most useful of these are the design tables presented by Cristal and Young[1]. These cover a variety of ripples, bandwidths, and sections, for coupling values of 3, 6, 8.34, 10.0, and 20.0 dB, and are extremely useful where optimum design for these specific coupling values are desired. General curves of normalized even-mode impedance for each of the various sections versus coupling values from 3.0 to 24.0 dB have been plotted for the 0.2 dB ripple case in Figures 5–8, 5–9, 5–10, and 5–11 for 3-, 5-, 7-, and 9-section couplers. Construction can be achieved by any of the means previously presented.

Like the symmetric coupler, the asymmetric coupler is built using a series of quarter-wave coupled sections. While the symmetric coupler has its tightest coupling section in the center, the asymmetric coupler has a tight coupling section at one end, decreasing with each section until the loosest coupling is reached at the opposite end. This is illustrated in Figure 5–12.

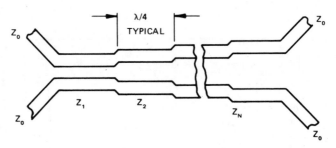

FIG. 5–12   Asymmetric Multi-Section Directional Coupler
Configuration

In general, greater bandwidth can be achieved for the same number of sections with an asymmetric coupler than with a symmetric coupler. The response will be an equal ripple response similar to that seen for the symmetric coupler and is illustrated in several examples of 10.0 dB couplers in Figure 5–13.

The first definitive paper on the synthesis of this classic coupler was presented by Levy.[2] In that paper, he presented amplitude response curves similar to those shown in Figure 5–13, but did not mention phase response, which is a major difference between the asymmetric and symmetric multi-section coupler. The symmetric coupler, regardless of the number of sections, has a nominal phase unbalance between the coupled arms of 90°. The asymmetric coupler has a

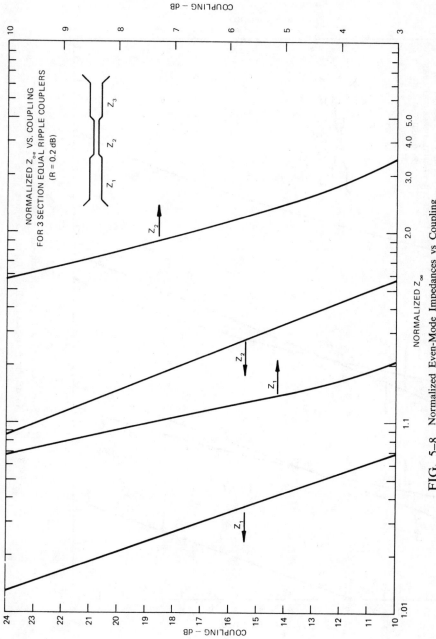

FIG. 5–8 Normalized Even-Mode Impedances vs Coupling in dB for Three-Section Equal-Ripple Couplers

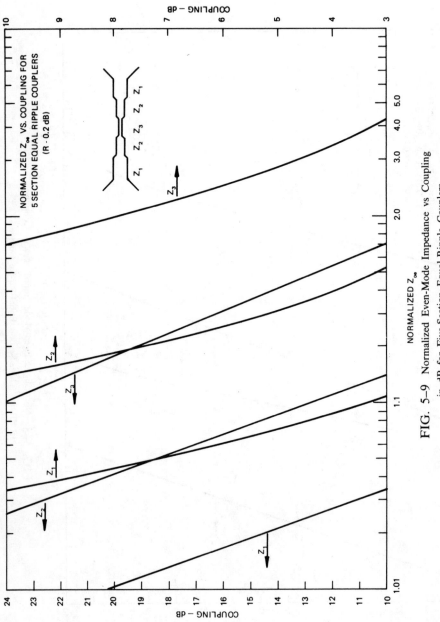

FIG. 5–9 Normalized Even-Mode Impedance vs Coupling in dB for Five-Section Equal-Ripple Couplers

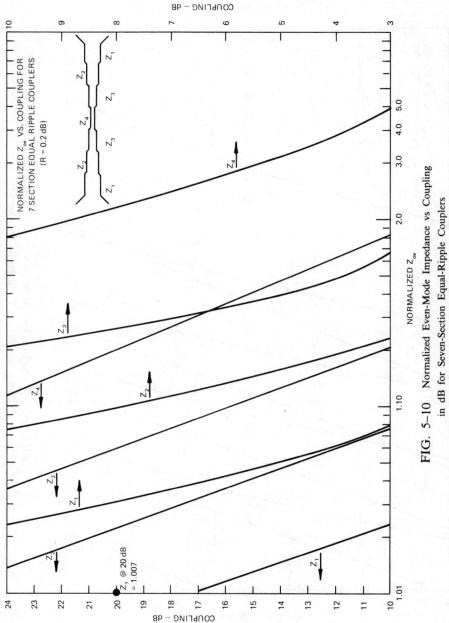

FIG. 5–10   Normalized Even-Mode Impedance vs Coupling in dB for Seven-Section Equal-Ripple Couplers

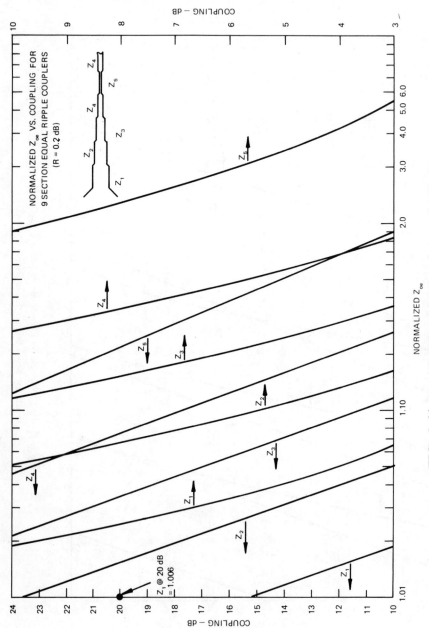

FIG. 5–11 Normalized Even-Mode Impedance vs Coupling in dB for Nine-Section Equal-Ripple Couplers

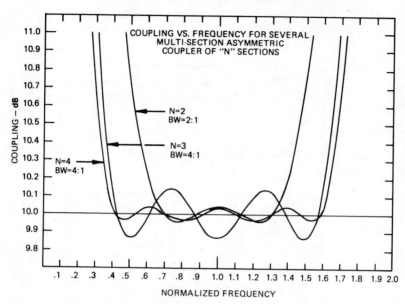

Figure 5–13 Frequency Response for Several Multi-Section
Asymmetric Couplers of 2, 3, and 4 Sections

varying phase response as a function of the number of sections. The two-section design has a 180° phase unbalance between the output arms at the center frequency and varies to each side of 180° as a function of frequency. A three-section device is centered at a nominal 90° phase response and varies from 0–180° over an octave. The four-section coupler has an in-phase response at the center frequency with a rapidly varying response as the result of its many sections. As a result, the asymmetric design cannot be used for circumstances where fixed phase response of the output coupled lines is required, although it does have application in a number of situations where the varying phase response can be used to perform some circuit function. It is also particularly useful, because of an increased bandwidth for its smaller number of sections, since it permits reduced size and lower loss.

Tables of coupling values for asymmetric couplers are available from several sources.[3][4] They are complete, and in many cases, go beyond the values which would normally be required for component design.

Tables 5–2 through 5–5 are a compilation of the normalized even-mode impedances for each of the sections of two-, three-, four-, and five-section asymmetrical directional couplers for amplitude ripples which the author considers to be of value in everyday design. For examples beyond these values, it is recommended that the original papers and tables be consulted.

		3 dB	6 dB	8.34 dB	10 dB	15 dB	20 dB
N = 2	Z_1	2.9655	1.9866	1.6409	1.5073	1.2533	1.1344
BW = 2:1	Z_2	1.2308	1.1481	1.1083	1.0882	1.0480	1.0266
N = 2	Z_1	3.1292	2.0596	1.6848	1.5412	1.2688	1.1423
BW = 3:1	Z_2	1.3206	1.2046	1.1489	1.1210	1.0656	1.0362
N = 2	Z_1	3.2895	2.1295	1.7261	1.5729	1.2833	1.1496
BW = 4:1	Z_2	1.4235	1.2680	1.1941	1.1573	1.0847	1.0466

TABLE 5-2

		3 dB	6 dB	8.34 dB	10 dB	15 dB	20 dB
N = 3	Z_1	3.5255	2.2239	1.7794	1.6140	1.3016	1.1578
BW = 4:1	Z_2	1.6386	1.3875	1.2751	1.2215	1.1174	1.0642
	Z_3	1.1382	1.0920	1.0683	1.0559	1.0308	1.0172
N = 3	Z_1	3.7672	2.3215	1.8346	1.6564	1.3203	1.1680
BW = 6:1	Z_2	1.8633	1.5112	1.3582	1.2870	1.1503	1.0816
	Z_3	1.2520	1.1661	1.1226	1.1000	1.0547	1.0303
N = 3	Z_1	3.9828	2.4058	1.8811	1.6921	1.3359	1.1756
BW = 8:1	Z_2	2.0742	1.6229	1.4315	1.3443	1.1785	1.0963
	Z_3	1.3801	1.2475	1.1813	1.1473	1.0798	1.0440

TABLE 5-3

		3 dB	6 dB	8.34 dB	10 dB	15 dB	20 dB
N = 4	Z_1	3.9210	2.3787	1.8654	1.6802	1.3306	1.1730
	Z_2	2.0471	1.6028	1.4168	1.3328	1.1726	1.0932
BW = 6:1	Z_3	1.3789	1.2415	1.1755	1.1424	1.0769	1.0424
	Z_4	1.1108	1.0749	1.0560	1.0459	1.0254	1.0142
N = 4	Z_1	4.2556	2.5048	1.9336	1.7326	1.3532	1.1841
	Z_2	2.4379	1.7961	1.5393	1.4279	1.2181	1.1168
BW = 10:1	Z_3	1.6554	1.4047	1.2890	1.2327	1.1235	1.0675
	Z_4	1.2753	1.1831	1.1355	1.1105	1.0604	1.0335

TABLE 5-4

		3 dB	6 dB	8.34 dB	10 dB	15 dB	20 dB
N = 5	Z_1	4.2052	2.4839	1.9218	1.7237	1.3494	1.1822
	Z_2	2.4053	1.7760	1.5256	1.4173	1.2129	1.1141
	Z_3	1.6410	1.3920	1.2776	1.2248	1.1193	1.0652
BW = 8:1	Z_4	1.2761	1.1800	1.1322	1.1077	1.0587	1.0325
	Z_5	1.0981	1.0668	1.0263	1.0411	1.0228	1.0127
	Z_1	4.4595	2.5762	1.9706	1.7612	1.3654	1.1900
	Z_2	2.7439	1.9326	1.6212	1.4911	1.2473	1.1316
N = 5	Z_3	1.9147	1.5423	1.3798	1.3040	1.1589	1.0861
	Z_4	1.4703	1.2995	1.2172	1.1759	1.0945	1.0520
BW = 12:1	Z_5	1.2206	1.1484	1.1104	1.0903	1.0496	1.0276

TABLE 5-5

Tandem Configurations

Whenever multi-section couplers are constructed, a significant factor controlling their practical realization is the center or tightest section coupling. In all cases, this will be significantly tighter than the overall value of coupling for the finished device. In the case of 3.0 dB couplers, in particular, this value of coupling frequently becomes unreasonable to manufacture either because the physical gap spacings are not realizable within the available ground plane spacing, or because the mechanical discontinuities of such a section are so great that the directivity of the coupler is severely degraded due to the interface mismatch. A method of minimizing this problem[5] has been presented. It consists of tandeming several lesser value couplers in order to create the final coupler. Two or more couplers may be tandemed, however, in most cases it has been shown that two is sufficient for coupling values in the order of 3 dB. This is illustrated in Figure 5–14a.

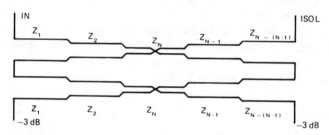

SYMMETRIC TANDEM OF SYMMETRIC 8.34 dB COUPLERS

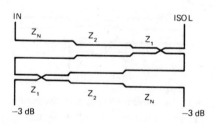

SYMMETRIC TANDEM OF ASYMMETRIC 8.34 dB COUPLERS

FIG. 5–14 Symmetric Tandem of Symmetric 8.34 dB Couplers and Symmetric Tandem of Asymmetric 8.34 dB Couplers

The value of coupling for each of the couplers in the tandem section for a 3.0 dB coupler is 8.34 dB, and it is for this reason that this value of coupling has been included in the various tables. For the symmetric case, the interconnection is obvious and the 90° phase quadrature characteristics of the coupler are maintained even in the tandem configuration.

For the asymmetric coupler, Hinden and Rosenzweig[6] have shown that there is only one proper interconnection of two asymmetric couplers in order to achieve a proper tandem. This is shown in Figure 5–14b. Like the single asymmetric coupler, the phase relationship of the tandem asymmetric coupler is not 90°, but varies with the number of sections as does the slope of the variation versus frequency. Nevertheless, with power dividers or directional couplers where it is desirable to have a broad bandwidth with a minimum number of sections and the coupling cannot be achieved in a single section asymmetric coupler, the tandem configuration has merit. As in the case of the symmetric tandem coupler, the proper coupler value for a 3.0 dB hybrid of each of the individual tandem couplers is 8.34 dB and couplers of this value have also been included in the design tables for asymmetric couplers. More complete tables for this specific value of coupling are available.[7]

In addition to the symmetric tandem of symmetric or non-symmetric couplers, there is also the configuration which consists of an asymmetric tandem of symmetric couplers.[8] This is illustrated by Figure 5–15, and as can be seen, consists of a tandem of symmetric couplers

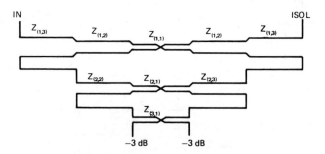

FIG. 5–15   Asymmetric Tandem of Symmetric Couplers
Creating a 3.0 dB Hybrid

each of which may have an unequal number of sections with respect to the others. While this configuration can be used for any value of coupling, it is most notably applied to the 3.0dB hybrid case. As with the symmetric tandem of symmetric couplers, the 90° phase relationship is maintained between the output ports. The

Z	BW = 2:1 R = .026	BW = 3:1 R = .15	BW = 4:1 R = .35	BW = 3:1 R = .02	BW = 4:1 R = .08	BW = 6:1 R = .27	BW = 4:1 R = .02	BW = 6:1 R = .10	BW = 8:1 R = .24
(1,1)	1.83434	1.83434	1.83434	1.83434	1.09434	2.09434	2.09434	2.09434	2.30314
(1,2)	1.12703	1.16819	1.21288	1.26301	1.33060	1.42859	1.41390	1.51506	1.62595
(1,3)				1.04382	1.06584	1.11305	1.11420	1.17365	1.23508
(1,4)							1.02184	1.04593	1.07553
(2,1)	1.54661	1.64220	1.74099	1.81680	1.69297	1.85219	1.81883	1.98066	1.94389

Z	BW = 6:1 R = .04	BW = 8:1 R = .11	BW = 10:1 R = .21	BW = 12:1 R = .35	BW = 14:1 R = .50	BW = 12:1 R = .21	BW = 14:1 R = .32	BW = 20:1 R = .47	BW = 25:1 R = .74
(1,1)	2.30314	2.30314	2.30314	2.09434	2.09434	1.83434	1.83434	2.09434	2.09434
(1,2)	1.60869	1.70226	1.78416	1.90926	1.98940	1.83434	1.83434	2.09434	2.09434
(1,3)	1.22836	1.29312	1.35343	1.41596	1.47151	1.45509	1.50657	1.71217	1.82215
(1,4)	1.07861	1.11654	1.15508	1.19502	1.23342	1.23402	1.27195	1.43153	1.51579
(1,5)	1.01971	1.03687	1.05730	1.08073	1.10520	1.11446	1.14183	1.26421	1.33328
(1,6)						1.04784	1.06548	1.15593	1.21023
(1,7)								1.08455	1.12504
(2,1)	1.91450	2.04717	2.16699	1.58220	1.61998	1.83434	1.83434	2.09434	2.09434
(2,2)						1.06829	1.11229	1.11770	1.20185
(3,1)					1.61998	1.60720	1.67705	1.48808	1.60639

TABLE 5-6

design technique is one of synthesis, and complete design tables for a wide variety of sections, bandwidths, and ripples for 3.0 dB as well as other values of coupling are available.[9] A selected group of values for the 3.0 dB case, which are most useful and which have normalized even-mode impedances within the normal range of construction practice are presented in Table 5–6, for design bandwidths varying from 2:1 to 25:1. The section impedance nomenclature refers to Figure 5–15.

Non-Uniform Line Couplers

Although extremely wide bandwidths are theoretically possible with multi-section design in practice, particularly at the higher frequency ranges, the many step discontinuities at the end of each of the sections causes a degradation in coupler performance. This is most noticeable in the area of directivity, although some increase in ripple of the coupling and insertion loss characteristics are also observed. To eliminate this problem, a non-uniform line construction based on the equal-ripple multiple section coupler was introduced by Tresselt in 1966.[10] Tresselt's method is far too mathematically complex to present here; briefly, it consists of integrating the prototype coupling function against the cosine integral and plotting the normalized even mode impedance versus the actual distance along the length of coupler. This results in a smoothly undulating coupling function when plotted against the distance from the edge of the coupler to the center. This is illustrated in Figure 5–16.

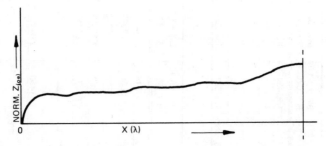

FIG. 5–16 Typical Plot of Normalized Even-Mode Impedances
vs Distance for a Symmetric
Non-Uniform Line Coupler

It also results in a coupler which is slightly longer ($\lambda/4$ at the design frequency) in coupled length than the equivalent equal-ripple multi-section prototype. The result is a highly effective technique for building wide-band high-directivity directional couplers, although it cannot be implemented without the use of a digital computer. Because of its immense value for high frequency broad-band couplers, tables are provided for the normalized even-mode impedances versus linear distance for coupling values of 3, 6, 8.34, 10, and 20 dB based on the Tresselt technique, using selected values for couplers from the Cristal and Young tables.[1] These are given in Tables 5–7a through 5–7e. The X-coordinate used in these tables is expressed in terms of fractions of a wavelength, starting at one edge of the coupler. The table covers 0 to 1.25 wavelengths, resulting in a coupler which is 2.5 wavelengths long where the wavelength is measured at the center frequency in the dielectric medium chosen. Figure 5–17 is a photograph of such a coupler.

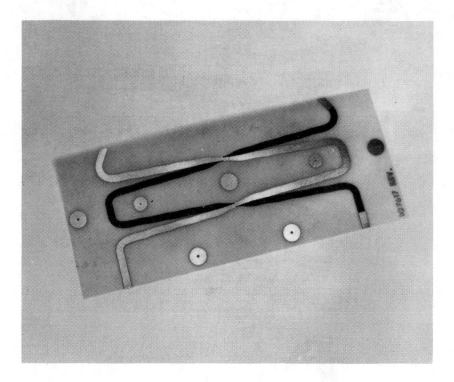

FIG. 5–17   Symmetric Non-Uniform Line Directional
Coupler

   Asymmetrical multi-section couplers may also be constructed by means of non-uniform line techniques. There is, however, a disadvantage to the asymmetric non-uniform line coupler because the tightly coupled end of the coupler has a major discontinuity between the input and output lines and the tightly coupled lines at the junction. This results in reduced directivity unless the junction is carefully matched. Figure 5–18 is a photograph of a coupler built using this technique.

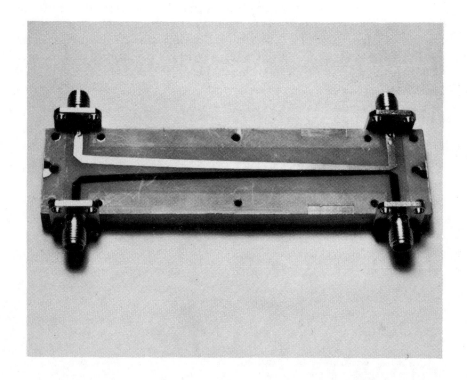

FIG. 5–18   Asymmetric Non-Uniform Line Directional
Coupler

3.01 DB COUPLER OF 9 EQUIVALENT SECTIONS
0.1 DB RIPPLE   9 :1 BANDWIDTH

X-COORD.	NORM. Z(OE)	X-COORD.	NORM. Z(OE)
0	1		
0.0125	1.00003	0.6375	1.21749
0.025	1.00024	0.65	1.23477
0.0375	1.00079	0.6625	1.25092
0.05	1.00182	0.675	1.26514
0.0625	1.00343	0.6875	1.27683
0.075	1.00566	0.7	1.28563
0.0875	1.00849	0.7125	1.2915
0.1	1.01187	0.725	1.29475
0.1125	1.01566	0.7375	1.29602
0.125	1.01971	0.75	1.29621
0.1375	1.02381	0.7625	1.29646
0.15	1.02778	0.775	1.29817
0.1625	1.03141	0.7875	1.3028
0.175	1.03456	0.8	1.31166
0.1875	1.0371	0.8125	1.3258
0.2	1.03898	0.825	1.34598
0.2125	1.04021	0.8375	1.37255
0.225	1.04089	0.85	1.40538
0.2375	1.04115	0.8625	1.44385
0.25	1.04119	0.875	1.4868
0.2625	1.04124	0.8875	1.53254
0.275	1.04163	0.9	1.57896
0.2875	1.04265	0.9125	1.62366
0.3	1.04456	0.925	1.66423
0.3125	1.04756	0.9375	1.69854
0.325	1.05174	0.95	1.72509
0.3375	1.05708	0.9625	1.74328
0.35	1.06347	0.975	1.75361
0.3625	1.07069	0.9875	1.75774
0.375	1.07845	1	1.75836
0.3875	1.08636	1.0125	1.75918
0.4	1.09406	1.025	1.76504
0.4125	1.10117	1.0375	1.78138
0.425	1.10735	1.05	1.8139
0.4375	1.11237	1.0625	1.8685
0.45	1.1161	1.075	1.95134
0.4625	1.11857	1.0875	2.06896
0.475	1.11993	1.1	2.22824
0.4875	1.12046	1.1125	2.43627
0.5	1.12053	1.125	2.69983
0.5125	1.12065	1.1375	3.02434
0.525	1.12143	1.15	3.41221
0.5375	1.12352	1.1625	3.8605
0.55	1.12748	1.175	4.35835
0.5625	1.13371	1.1875	4.88465
0.575	1.14246	1.2	5.40721
0.5875	1.15375	1.2125	5.88442
0.6	1.16739	1.225	6.27028
0.6125	1.183	1.2375	6.52221
0.625	1.19995	1.25	6.60985

TABLE 5-7 a

6 DB COUPLER OF 9 EQUIVALENT SECTIONS
0.1 DB RIPPLE   8 :1 BANDWIDTH

X-COORD.	NORM. Z(OE)	X-COORD.	NORM. Z(OE)
0	1		
0.0125	1.00002	0.6375	1.12741
0.025	1.00013	0.65	1.13763
0.0375	1.00043	0.6625	1.14714
0.05	1.00098	0.675	1.15548
0.0625	1.00185	0.6875	1.16231
0.075	1.00305	0.7	1.16743
0.0875	1.00458	0.7125	1.17084
0.1	1.00639	0.725	1.17273
0.1125	1.00843	0.7375	1.17347
0.125	1.0106	0.75	1.17358
0.1375	1.0128	0.7625	1.17372
0.15	1.01491	0.775	1.17474
0.1625	1.01685	0.7875	1.17749
0.175	1.01852	0.8	1.18272
0.1875	1.01987	0.8125	1.19107
0.2	1.02087	0.825	1.20292
0.2125	1.02153	0.8375	1.21843
0.225	1.02189	0.85	1.23746
0.2375	1.02203	0.8625	1.25956
0.25	1.02205	0.875	1.28399
0.2625	1.02208	0.8875	1.30975
0.275	1.02231	0.9	1.33562
0.2875	1.02292	0.9125	1.36028
0.3	1.02407	0.925	1.38247
0.3125	1.02587	0.9375	1.40109
0.325	1.02837	0.95	1.4154
0.3375	1.03157	0.9625	1.42517
0.35	1.03538	0.975	1.4307
0.3625	1.03968	0.9875	1.4329
0.375	1.04428	1	1.43324
0.3875	1.04897	1.0125	1.43365
0.4	1.05352	1.025	1.43667
0.4125	1.0577	1.0375	1.44505
0.425	1.06133	1.05	1.46165
0.4375	1.06427	1.0625	1.48927
0.45	1.06646	1.075	1.53062
0.4625	1.0679	1.0875	1.58824
0.475	1.06869	1.1	1.66436
0.4875	1.069	1.1125	1.76084
0.5	1.06905	1.125	1.87881
0.5125	1.06911	1.1375	2.01838
0.525	1.06959	1.15	2.17814
0.5375	1.07087	1.1625	2.35466
0.55	1.07329	1.175	2.54205
0.5625	1.07708	1.1875	2.73175
0.575	1.0824	1.2	2.91277
0.5875	1.08924	1.2125	3.07252
0.6	1.09747	1.225	3.19822
0.6125	1.10684	1.2375	3.27874
0.625	1.11698	1.25	3.30649

TABLE 5-7 b

8.34 DB COUPLER OF 9 EQUIVALENT SECTIONS
0.1 DB RIPPLE   7.7 :1 BANDWIDTH

X-COORD.	NORM. Z(OE)	X-COORD.	NORM. Z(OE)
0	1		
0.0125	1.00001	0.6375	1.09114
0.025	1.00009	0.65	1.09843
0.0375	1.0003	0.6625	1.1052
0.05	1.00069	0.675	1.11112
0.0625	1.0013	0.6875	1.11597
0.075	1.00214	0.7	1.11959
0.0875	1.0032	0.7125	1.12201
0.1	1.00447	0.725	1.12335
0.1125	1.0059	0.7375	1.12387
0.125	1.00741	0.75	1.12394
0.1375	1.00894	0.7625	1.12405
0.15	1.01042	0.775	1.12477
0.1625	1.01177	0.7875	1.12672
0.175	1.01293	0.8	1.13045
0.1875	1.01387	0.8125	1.13637
0.2	1.01457	0.825	1.14477
0.2125	1.01503	0.8375	1.15572
0.225	1.01527	0.85	1.16911
0.2375	1.01537	0.8625	1.18459
0.25	1.01539	0.875	1.20163
0.2625	1.01541	0.8875	1.2195
0.275	1.01558	0.9	1.23736
0.2875	1.01602	0.9125	1.2543
0.3	1.01685	0.925	1.26948
0.3125	1.01815	0.9375	1.28216
0.325	1.01996	0.95	1.29189
0.3375	1.02226	0.9625	1.29851
0.35	1.02501	0.975	1.30225
0.3625	1.02811	0.9875	1.30374
0.375	1.03141	1	1.30397
0.3875	1.03478	1.0125	1.30425
0.4	1.03804	1.025	1.30626
0.4125	1.04104	1.0375	1.31185
0.425	1.04364	1.05	1.3229
0.4375	1.04574	1.0625	1.34121
0.45	1.04731	1.075	1.36846
0.4625	1.04834	1.0875	1.4061
0.475	1.04891	1.1	1.45529
0.4875	1.04913	1.1125	1.51677
0.5	1.04916	1.125	1.59074
0.5125	1.04921	1.1375	1.67669
0.525	1.04955	1.15	1.77316
0.5375	1.05048	1.1625	1.87759
0.55	1.05223	1.175	1.98619
0.5625	1.05497	1.1875	2.09399
0.575	1.05881	1.2	2.19501
0.5875	1.06375	1.2125	2.28279
0.6	1.06968	1.225	2.351
0.6125	1.07641	1.2375	2.39433
0.625	1.08368	1.25	2.40919

TABLE 5-7 c

10 DB COUPLER OF 9 EQUIVALENT SECTIONS
0.2 DB RIPPLE 9.3 :1 BANDWIDTH

X-COORD.	NORM. Z(OE)	X-COORD.	NORM. Z(OE)
0	1		
0.0125	1.00001	0.6375	1.08895
0.025	1.00011	0.65	1.0954
0.0375	1.00037	0.6625	1.10138
0.05	1.00084	0.675	1.10661
0.0625	1.00159	0.6875	1.11088
0.075	1.00262	0.7	1.11408
0.0875	1.00394	0.7125	1.11621
0.1	1.00549	0.725	1.11739
0.1125	1.00724	0.7375	1.11785
0.125	1.0091	0.75	1.11792
0.1375	1.01099	0.7625	1.118
0.15	1.01281	0.775	1.11861
0.1625	1.01447	0.7875	1.12024
0.175	1.0159	0.8	1.12336
0.1875	1.01706	0.8125	1.12831
0.2	1.01792	0.825	1.13532
0.2125	1.01848	0.8375	1.14445
0.225	1.01879	0.85	1.15559
0.2375	1.0189	0.8625	1.16846
0.25	1.01892	0.875	1.18258
0.2625	1.01894	0.8875	1.19736
0.275	1.01911	0.9	1.2121
0.2875	1.01954	0.9125	1.22605
0.3	1.02035	0.925	1.23852
0.3125	1.02161	0.9375	1.24892
0.325	1.02336	0.95	1.25689
0.3375	1.0256	0.9625	1.2623
0.35	1.02828	0.975	1.26536
0.3625	1.03128	0.9875	1.26658
0.375	1.0345	1	1.26677
0.3875	1.03777	1.0125	1.26699
0.4	1.04094	1.025	1.26859
0.4125	1.04385	1.0375	1.27302
0.425	1.04637	1.05	1.28176
0.4375	1.04842	1.0625	1.29622
0.45	1.04994	1.075	1.31766
0.4625	1.05094	1.0875	1.34716
0.475	1.05149	1.1	1.38549
0.4875	1.0517	1.1125	1.43307
0.5	1.05173	1.125	1.48985
0.5125	1.05178	1.1375	1.55521
0.525	1.05208	1.15	1.62785
0.5375	1.05291	1.1625	1.70566
0.55	1.05446	1.175	1.78575
0.5625	1.0569	1.1875	1.86444
0.575	1.06031	1.2	1.93752
0.5875	1.06468	1.2125	2.00051
0.6	1.06994	1.225	2.04915
0.6125	1.07591	1.2375	2.07992
0.625	1.08235	1.25	2.09044

TABLE 5-7 d

20 DB COUPLER OF 9 EQUIVALENT SECTIONS
0.2 DB RIPPLE 9 :1 BANDWIDTH

X-COORD.	NORM. Z(OE)	X-COORD.	NORM. Z(OE)
0	1		
0.0125	1.	0.6375	1.02627
0.025	1.00003	0.65	1.02814
0.0375	1.00011	0.6625	1.02987
0.05	1.00025	0.675	1.03137
0.0625	1.00047	0.6875	1.0326
0.075	1.00077	0.7	1.03351
0.0875	1.00116	0.7125	1.03412
0.1	1.00162	0.725	1.03446
0.1125	1.00214	0.7375	1.03459
0.125	1.00268	0.75	1.03461
0.1375	1.00324	0.7625	1.03463
0.15	1.00377	0.775	1.03481
0.1625	1.00426	0.7875	1.03528
0.175	1.00468	0.8	1.03617
0.1875	1.00502	0.8125	1.03758
0.2	1.00527	0.825	1.03957
0.2125	1.00543	0.8375	1.04216
0.225	1.00552	0.85	1.04529
0.2375	1.00555	0.8625	1.04888
0.25	1.00556	0.875	1.0528
0.2625	1.00557	0.8875	1.05686
0.275	1.00561	0.9	1.06087
0.2875	1.00574	0.9125	1.06464
0.3	1.00599	0.925	1.06798
0.3125	1.00637	0.9375	1.07075
0.325	1.00689	0.95	1.07287
0.3375	1.00757	0.9625	1.0743
0.35	1.00837	0.975	1.0751
0.3625	1.00926	0.9875	1.07542
0.375	1.01022	1	1.07547
0.3875	1.0112	1.0125	1.07553
0.4	1.01214	1.025	1.07595
0.4125	1.013	1.0375	1.0771
0.425	1.01374	1.05	1.07936
0.4375	1.01435	1.0625	1.08309
0.45	1.0148	1.075	1.08856
0.4625	1.01509	1.0875	1.09599
0.475	1.01525	1.1	1.10548
0.4875	1.01532	1.1125	1.117
0.5	1.01533	1.125	1.13041
0.5125	1.01534	1.1375	1.14542
0.525	1.01543	1.15	1.16159
0.5375	1.01567	1.1625	1.17837
0.55	1.01614	1.175	1.1951
0.5625	1.01686	1.1875	1.21103
0.575	1.01787	1.2	1.22542
0.5875	1.01916	1.2125	1.23752
0.6	1.02071	1.225	1.24669
0.6125	1.02247	1.2375	1.25241
0.625	1.02435	1.25	1.25435

TABLE 5-7 e

REFERENCES

1   Cristal, E. G., and Young, L., *Theory and Tables of Optimum Symmetrical TEM Mode Coupled Transmission Line Directional Couplers,* MTT-13, No. 5, September 1965, pp. 544–558.

2   Levy, R., *General Synthesis of Asymmetric Multi-Element Directional Couplers,* MTT-11, No. 4, July 1963, pp. 226–237.

3   Levy, R., *Tables for Asymmetric Multi-Element Coupled Transmission Line Directional Couplers,* MTT-12, No. 3, May 1964, pp. 275–279.

4   Hinden, H. J., and Rosenzweig, A., *3 dB Couplers Constructed From Two Tandem Connected 8.34 dB Asymmetric Couplers,* MTT-16, No. 2, February 1968, pp.125–126.

5   Shelton, R., Wolf, J., and van Wagoner, R., *Tandem Couplers and Phase Shifters For Multi-Octave Bandwidth,* Microwaves, April 1965, pp. 14–19.

6   Hinden and Rosenzweig, op. cit.

7   ibid.

8   Shelton, P., and Mosko, J., *Synthesis and Design of Wideband Equal Ripple TEM Directional Couplers and Fixed Phase Shifters,* MTT-14, No. 10, October 1966, pp. 462–473.

9   Shelton, P., and Mosko, J., *Design Tables for Directional Couplers and Phase Shifters,* Library of Congress, Photo Duplication Service Document No. 9017.

10   Tresselt, C. P., *The Design and Construction of Broad Band High Directivity 90° Couplers Using Non-Uniform Line Techniques,* MTT-14, No. 12, December 1966 pp. 647–656.

11   Arndt, F., *Tables for Asymmetric Tchebyshev High-Pass TEM Mode Directional Couplers,* MTT-18, No. 9, September 1970, pp. 633–638.

CHAPTER

—6

Low Pass and High Pass Filters

In the construction of any microwave circuitry, it is frequently desirable to have frequency selectivity. This may be in the form of low-pass filtering, high-pass filtering, bandpass filtering, or band-stop filtering. These subjects will be discussed in Chapters 6 and 7. The subject of microwave filters is extensive and has been widely covered in the literature. Neither chapter will go deeply into the subject of filters but will present design procedures and curves for those filters which are most readily adaptable to printed stripline construction. Thus, only a small segment of the available classes of microwave filters will be covered and there will be no reference to the modern methods of filter synthesis.

Low-Pass Filters

There are three general categories of low-pass filters in common stripline usage. All of them are based on translations, by means of short sections of transmission line, from lumped-constant prototypes; thus, because of the limitations of these transmission-line translations, they are, less than perfect in their response when compared to the theoretical prototype.

Perhaps one of the most widely used is the constant-K, m-derived end section type of low-pass filter. The constant-K reactance network is one in which there are inverse reactance arms, shown in Equation 6-1.

$$Z_1 Z_2 = K^2 = R^2 \tag{6-1}$$

181

Since K^2 is independent of frequency, we have equations 6–2 and 6–3.

$$Z_1 = j\omega L = R\sqrt{1 - (\omega L/2R)^2} \tag{6-2}$$

$$Z_2 = \frac{1}{j\omega C} \tag{6-3}$$

Inasmuch as this class of filter has a relatively low roll-off versus frequency, it is normally desirable to add m-derived end sections where m is chosen to provide a pole of attenuation near the cutoff frequency as well as to provide flat terminal-image impedance for the filter. The impedances of the end sections will be as defined in equation 6–4 and 6–5; with m equal to equation 6–6.

$$Z_{1m} = m\,Z_1 K \tag{6-4}$$

$$Z_{2m} = \frac{1 - m^2}{4\,m}\;Z_1 K + \frac{Z_2 K}{m} \tag{6-5}$$

$$m = \sqrt{1 - \left(\frac{f_0}{f^\infty}\right)^2} \tag{6-6}$$

Although it would appear from equation 6–6 that the value of m can be any reasonable number, in practice it has been found that the value 0.6 for m provides the best in-band VSWR performance for filters of this type. This results in a pole of attenuation at 1.25 times the cutoff frequency of the filter.

The basic schematic for such a filter is shown in Figure 6–1 and its printed circuit equivalent in Figure 6–2, which also defines the parameters within the design procedure.

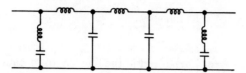

FIG. 6–1   General Schematic for a Lumped-Element,
Constant-K, m-derived End Section,
Low-Pass Filter

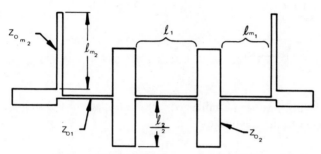

FIG. 6–2   Stripline Realization of a Constant-K,
m-derived End Section, Low-Pass Filter
Showing Design Dimensions

This design procedure is well established and has been proven in a wide variety of applications. The high impedance lines Z_{01} should be chosen to be as high as practical in the type of construction to be used, in order to minimize the line length. This will increase the frequency of the second response. Once Z_{01} has been chosen, the length ℓ_1 can be calculated from equations 6–7, 6–8, and 6–9.

$$\ell_1 = \frac{V}{\omega} \ \text{arc tan} \ \frac{Z \ (Z_o)}{Z_{01}} \tag{6–7}$$

$$V = \frac{1.1803 \times 10^{10}}{\sqrt{\epsilon}} \tag{6–8}$$

$$\omega = 2\pi F_{(GHz)} \times 10^9 = \text{Radians/sec.} \tag{6-9}$$

Although there is an error introduced in the line lengths because of the junction capacities and the effect of fringing capacity at the ends of the shunt arms, corrections have not been included since, unlike band-pass filters, the skirt response of this type of filter is usually not critical. The errors are small and overshadowed by the response of the m-derived end section. Similarly, the shunt low-impedance section may be calculated from equation 6–10 where λ_c is equal to the wavelength of the frequency of cutoff and λ max is equal to the wavelength at the highest frequency of the stop-band attenuation.

$$\frac{\ell_2}{2} = \frac{\sqrt{\lambda c \cdot \lambda max.}}{4} \tag{6-10}$$

Once this length has been established, the impedance of the shunt section may be established from equation 6–11.

$$Z_{02} = Z_0 \left(\tan \frac{\omega \, \ell_2}{V} \right) \tag{6-11}$$

The length of the series line in the m-derived end section may be calculated from equation 6–12 making use of the same value of high impedance line as used in the calculation of ℓ_{o1}. The shunt arm on the m-derived end section will be a quarter wavelength long at the frequency of infinite attenuation, as established in equations 6–13 and 6–14. Its impedance may then be calculated from equation 6–15.

$$\ell m_1 = \frac{V}{\omega} \arctan \left[\left(\frac{m+1}{Z_{0_1}} \right) Z_0 \right] \tag{6-12}$$

$$\ell m_2 = \frac{\lambda_{f\infty}}{4} \tag{6-13}$$

$$\text{Where } f_\infty = \frac{f_0}{\sqrt{1-m^2}} \tag{6-14}$$

$$Z_{0_{m_2}} = m (Z_0) \tan \frac{\omega \ell\, m_2}{V} \qquad\qquad (6\text{--}15)$$

For the recommended value of m = 0.6, this will result in a class of filters whose frequency response is shown in Figure 6–3. These curves represent the attenuation versus normalized frequency for theoretically perfect lumped-constant low-pass filters having a varying number of central constant-K sections. At lower frequencies they can be realized perfectly with lumped-constant elements.

The printed version of the filters which have just been described will not, however, be as perfect as the lumped-constant prototypes. Degradation will occur as a result of the imperfect substitution of short-line sections for the lumped elements. This imperfection shows up as a higher insertion loss at the point of cutoff and a lightly shallower slope on the attenuation curves. This is illustrated in Figure 6–4. Both figures show the pole of attenuation which occurs at 1.25 times the frequency of cutoff. Since it is not always necessary to have this sharp skirt, but merely to have a high point of attenuation at some point above the frequency of operation, it is often desirable to set the cutoff of a low-pass filter so that the pole of highest attenuation occurs at the frequency which requires the highest stop-band attenuation.

This will have the effect of minimizing the number of sections needed. Although in general it will be desirable to have the highest stop-band frequency limit possible, there may be limitations in shunt-stub impedance which make this impractical. Figure 6–5 relates the shunt-stub impedance and shunt-stub length versus the upper stop-band limit, and may be used to optimize the design for a particular application. In similar fashion, Figure 6–6 describes the relationship between L_1 and L_m and their line impednaces. It will be noted that the impedance and length of the shunt stub on the m-derived end section remains constant for m = 0.6. Although this type of filter is probably the most widely used stripline low-pass filter, it does take more surface area than some of the other types of construction and so, unless the immediate high attenuation pole in the stop band is required, it may become desirable to use other circuit configurations.

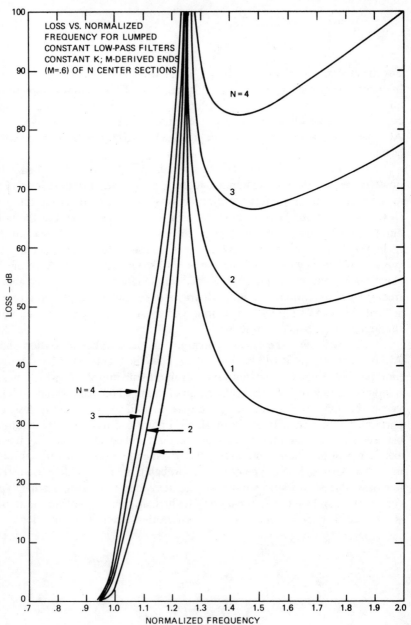

FIG. 6–3 Theoretical Response for Lumped-Constant,
Low-Pass Filters with a Constant-K,
m-derived End Section, where m = 0.6 for
1, 2, 3, and 4 Center Sections

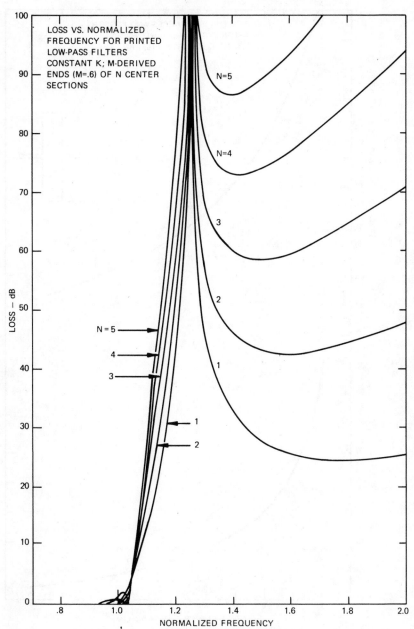

FIG. 6–4 Theoretical Response for Printed Low-Pass Filters
with a Constant-K, m-derived End Section,
where m = 0.6 for 1, 2, 3, 4, and 5
Center Sections

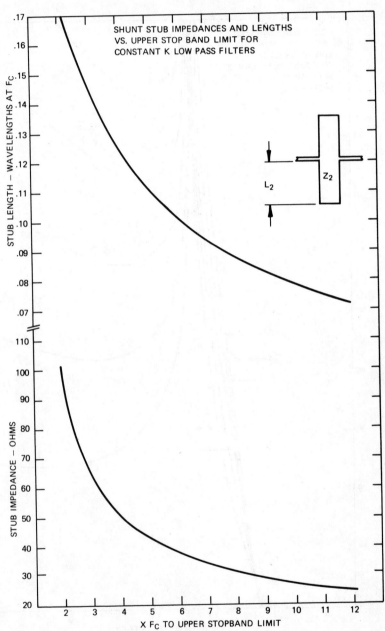

FIG. 6–5 Shunt Stub Impedances and Line Lengths vs. Upper
Stop Band Limit for Constant-K, Low-Pass
Filters

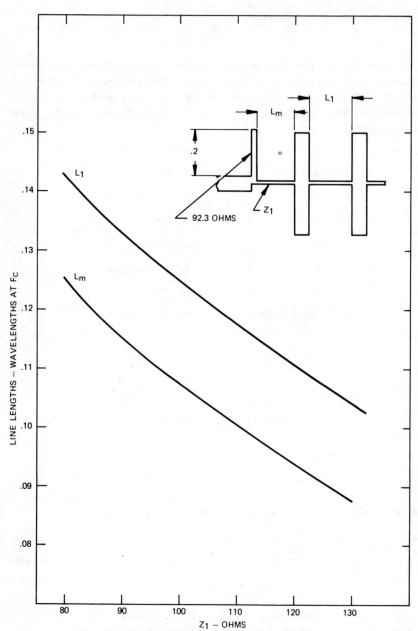

FIG. 6–6 Line Lengths vs Characteristic Impedance for
Series-Line Sections for Low-Pass Filters
of the Constant **K**, m-derived End Section
Type, where m = 0.6

High-Z/Low-Z Low-Pass Filters

A more classic type of low-pass filter is the series-inductance/shunt-capacitance type of filter shown in Figure 6–7a. It can be designed with either a maximally flat frequency response or an equal-ripple Tchebyshev response in the pass-band Its design is based upon the normalized capacitance and inductance values or G-values.

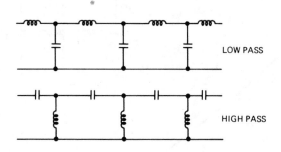

FIG. 6–7   General Schematic for Lumped-Element Low-Pass
and High-Pass Ladder Filters

If we assume for our purposes that $G_0 = 1$, and then

$$g_1 = g_{(N+1)} = 1 \tag{6–16}$$

for the maximally flat case, G_k will be described by

$$g_K = 2 \operatorname{Sin} \left[\frac{(2K-1)\pi}{2N} \right] \tag{6–17}$$

$$K = 1, 2, \ldots \ldots N$$

for values of k from 1 to n, where n is the number of sections in the filter.

The frequency response of this class of filter can be accurately predicted and is shown in Figure 6–8 for two to fifteen sections. By definition, a normalized frequency equal to 1 occurs at the 3 dB cutoff of a maximally flat filter.

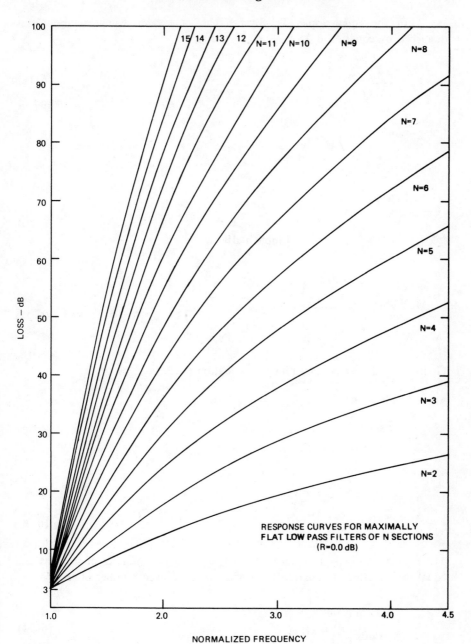

FIG. 6–8 Frequency Response Curves for Maximally Flat,
Low-Pass Filters having
two to fifteen Sections

For the equal-ripple, Tchebyshev case, G_0 again equals 0 and G_1 is described by equation 6–18.

$$g_1 = \frac{2 a_1}{\gamma} \tag{6-18}$$

$$\gamma = \sinh\left(\frac{\beta}{2N}\right) \tag{6-19}$$

$$\beta = \ell n\left(\text{Coth}\ \frac{R}{17.37}\right) \tag{6-20}$$

$$R = \text{Ripple in dB}$$

$$a_K = \sin\left[\frac{(2K-1)\pi}{2N}\right] \tag{6-21}$$

$$K = 1, 2, \ldots\ldots N$$

G_k for k = 2 to n is determined by equation 6–22.

$$g_K = \frac{4a_{K-1}\, a_K}{b_{K-1} g_{K-1}} \tag{6-22}$$

$$K = 2, 3, \ldots\ldots\ldots N$$

$$b_K = \gamma^2 + \sin^2\left(\frac{K\pi}{N}\right) \tag{6-23}$$

$$K = 1, 2, \ldots\ldots\ldots N$$

The end-section G_{n+1} is equal to 1 when n = an odd number of sections and equal to the value of equation 6–24 when n = an even number of sections.

$$g_{(N+1)} = 1 \text{ for N odd}$$
$$= \text{Coth}^2\left(\frac{\beta}{4}\right) \text{ for N even} \tag{6-24}$$

The frequency response for this type of filter for ripples equal to 0.01 dB and 0.1 dB are shown in Figures 6–9 and 6–10. While it is, of course, possible to construct filters with a higher ripple value (which would result in sharper skirts), this is not generally an acceptable practice, inasmuch as the ripple is a reflective function which results in a higher VSWR. The 0.01 dB ripple results in a 1.1 VSWR while the 0.1 dB ripple causes a 1.36 VSWR. Inasmuch as actual performance is generally somewhat worse than the theoretical predictions, curves for the higher values of ripple have not been included. They are, however, readily available.[2] Tables of the G-values necessary to construct filters of this type are given in Tables 6–1, 6–2, and 6–3, for the maximally flat 0 dB ripple case, as well as for the 0.01 and 0.1 cases. It should be mentioned that these G-value tables provide the cornerstone, not only of the low-pass and high-pass in Chapter 6, but also high-pass filters and the bandpass and band-stop filters in Chapter 7.

The normalized G-value must be converted into inductance or capacitance to be used in the form of Figure 6–7 by means of equations 6–25 and 6–26, making use of the appropriate G-value for each section; thus for the filters described, each odd-number G-value will be a series inductance and each even-number G-value will be a shunt capacitance.

$$L(K) = \frac{G(K) \; Z_0}{2 \pi F} \quad \text{(Henries)} \qquad\qquad (6\text{-}25)$$

$$C(K) = \frac{G(K)}{2 \, Z_0 \; \pi F} \quad \text{(Farads)}$$
$$(6\text{-}26)$$
$$F = C.P.S. = \text{Hertz}$$

In order to maintain the input and output impedances at equal values, the class of filter as described herein should be built only for odd numbers of sections, i.e., $n = 3, 5, 7, 9$, etc., using the equations given. Even-numbered-section filters can be built for the maximally flat case with some modification.

In printed form this type of filter will be realized as a series of high-impedance/low impedance sections as shown in Figure 6–11. The length of the high-impedance inductive sections may be calculated by equation 6–27 where Z_1 is the impedance of the inductive section.

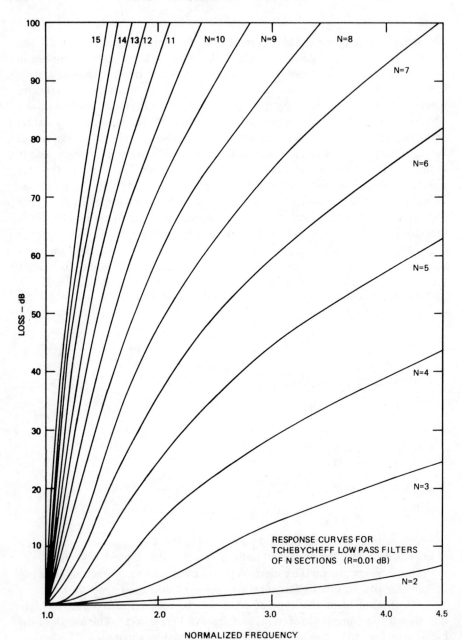

NORMALIZED FREQUENCY

FIG. 6–9   Frequency Response Curves for Tchebyshev
Low-Pass Filters having Ripple = 0.01 dB
for two to fifteen Sections

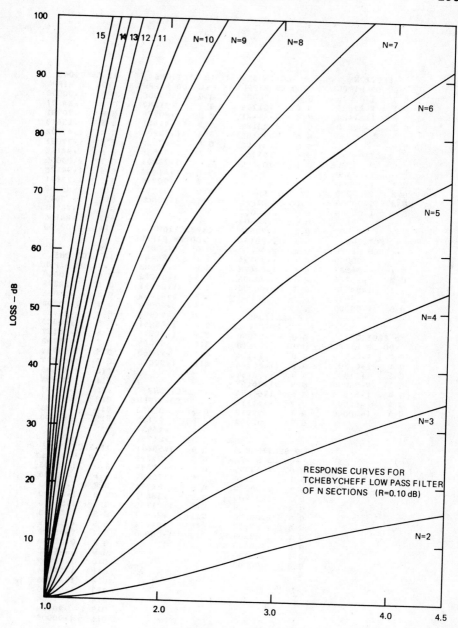

NORMALIZED FREQUENCY

FIG. 6–10 Frequency Response Curves for Tchebyshev
Low-Pass Filters having Ripple = 0.1 dB
and two to fifteen Sections

```
2 SECTIONS          6 SECTIONS          10 SECTIONS         13 SECTIONS
 •00 DB RIPPLE       •00 DB RIPPLE       •00 DB RIPPLE       •00 DB RIPPLE
G 0    1•00000      G 0    1•00000      G 0    1•00000      G 0    1•00000
G 1    1•41421      G 1     •51764      G 1     •31287      G 1     •24107
G 2    1•41422      G 2    1•41421      G 2     •90798      G 2     •70921
G 3    1•00000      G 3    1•93185      G 3    1•41421      G 3    1•13613
                    G 4    1•93185      G 4    1•78201      G 4    1•49702
                    G 5    1•41422      G 5    1•97538      G 5    1•77091
                    G 6     •51764      G 6    1•97538      G 6    1•94188
 3 SECTIONS         G 7    1•00000      G 7    1•78201      G 7    2•00000
 •00 DB RIPPLE                          G 8    1•41422      G 8    1•94188
G 0    1•00000                          G 9     •90799      G 9    1•77091
G 1    1•00000                          G10     •31287      G10    1•49702
G 2    2•00000       7 SECTIONS         G11    1•00000      G11    1•13613
G 3    1•00000       •00 DB RIPPLE                          G12     •70921
G 4    1•00000      G 0    1•00000                          G13     •24105
                    G 1     •44504                          G14    1•00000
                    G 2    1•24698       11 SECTIONS
                    G 3    1•80194       •00 DB RIPPLE
 4 SECTIONS         G 4    2•00000      G 0    1•00000       14 SECTIONS
 •00 DB RIPPLE      G 5    1•80194      G 1     •28463       •00 DB RIPPLE
G 0    1•00000      G 6    1•24698      G 2     •83083      G 0    1•00000
G 1     •76537      G 7     •44505      G 3    1•30972      G 1     •22393
G 2    1•84776      G 8    1•00000      G 4    1•68251      G 2     •66056
G 3    1•84776                          G 5    1•91899      G 3    1•06406
G 4     •76537                          G 6    2•00000      G 4    1•41421
G 5    1•00000                          G 7    1•91899      G 5    1•69345
                     8 SECTIONS         G 8    1•68251      G 6    1•88777
                     •00 DB RIPPLE      G 9    1•30972      G 7    1•98742
                    G 0    1•00000      G10     •83083      G 8    1•98742
                    G 1     •39018      G11     •28463      G 9    1•88777
 5 SECTIONS         G 2    1•11114      G12    1•00000      G10    1•69345
 •00 DB RIPPLE      G 3    1•66294                          G11    1•41422
G 0    1•00000      G 4    1•96157                          G12    1•06407
G 1     •61803      G 5    1•96157                          G13     •66056
G 2    1•61803      G 6    1•66294       12 SECTIONS        G14     •22393
G 3    2•00000      G 7    1•11114       •00 DB RIPPLE      G15    1•00000
G 4    1•61804      G 8     •39019      G 0    1•00000
G 5     •61804      G 9    1•00000      G 1     •26105
G 6    1•00000                          G 2     •76537
                                        G 3    1•21752       15 SECTIONS
                                        G 4    1•58671       •00 DB RIPPLE
                     9 SECTIONS         G 5    1•84776      G 0    1•00000
                     •00 DB RIPPLE      G 6    1•98289      G 1     •20906
                    G 0    1•00000      G 7    1•98289      G 2     •61803
                    G 1     •34730      G 8    1•84776      G 3    1•00000
                    G 2    1•00000      G 9    1•58671      G 4    1•33826
                    G 3    1•53209      G10    1•21753      G 5    1•61803
                    G 4    1•87938      G11     •76537      G 6    1•82709
                    G 5    2•00000      G12     •26106      G 7    1•95629
                    G 6    1•87939      G13    1•00000      G 8    2•00000
                    G 7    1•53209                          G 9    1•95630
                    G 8    1•00000                          G10    1•82709
                    G 9     •34730                          G11    1•61804
                    G10    1•00000                          G12    1•33826
                                                            G13    1•00000
                                                            G14     •61804
                                                            G15     •20906
                                                            G16    1•00000
```

2 SECTIONS
.01 DB RIPPLE
G 0 1.00000
G 1 .44889
G 2 .40781
G 3 1.10075

3 SECTIONS
.01 DB RIPPLE
G 0 1.00000
G 1 .62919
G 2 .97089
G 3 .62919
G 4 1.00000

4 SECTIONS
.01 DB RIPPLE
G 0 1.00000
G 1 .71288
G 2 1.20036
G 3 1.32130
G 4 .64763
G 5 1.10075

5 SECTIONS
.01 DB RIPPLE
G 0 1.00000
G 1 .75634
G 2 1.30493
G 3 1.57732
G 4 1.30493
G 5 .75635
G 6 1.00000

6 SECTIONS
.01 DB RIPPLE
G 0 1.00000
G 1 .78136
G 2 1.36002
G 3 1.68969
G 4 1.53503
G 5 1.49704
G 6 .70985
G 7 1.10075

7 SECTIONS
.01 DB RIPPLE
G 0 1.00000
G 1 .79696
G 2 1.39243
G 3 1.74814
G 4 1.63313
G 5 1.74814
G 6 1.39243
G 7 .79696
G 8 1.00000

8 SECTIONS
.01 DB RIPPLE
G 0 1.00000
G 1 .80789
G 2 1.41309
G 3 1.78245
G 4 1.68334
G 5 1.85294
G 6 1.61930
G 7 1.55546
G 8 .73341
G 9 1.10075

9 SECTIONS
.01 DB RIPPLE
G 0 1.00000
G 1 .81447
G 2 1.42706
G 3 1.80437
G 4 1.71254
G 5 1.90580
G 6 1.71254
G 7 1.80437
G 8 1.42706
G 9 .81448
G10 1.00000

10 SECTIONS
.01 DB RIPPLE
G 0 1.00000
G 1 .81966
G 2 1.43696
G 3 1.81927
G 4 1.73112
G 5 1.93624
G 6 1.75902
G 7 1.90553
G 8 1.65275
G 9 1.58174
G10 .74465
G11 1.10075

11 SECTIONS
.01 DB RIPPLE
G 0 1.00000
G 1 .82353
G 2 1.44423
G 3 1.82988
G 4 1.74372
G 5 1.95548
G 6 1.78555
G 7 1.95548
G 8 1.74372
G 9 1.82989
G10 1.44423
G11 .82354
G12 1.00000

12 SECTIONS
.01 DB RIPPLE
G 0 1.00000
G 1 .82648
G 2 1.44973
G 3 1.83773
G 4 1.75271
G 5 1.96849
G 6 1.80220
G 7 1.98378
G 8 1.78831
G 9 1.92930
G10 1.66952
G11 1.59579
G12 .75085
G13 1.10075

13 SECTIONS
.01 DB RIPPLE
G 0 1.00000
G 1 .82879
G 2 1.45398
G 3 1.84370
G 4 1.75936
G 5 1.97773
G 6 1.81341
G 7 2.00144
G 8 1.81341
G 9 1.97774
G10 1.75936
G11 1.84370
G12 1.45399
G13 .82881
G14 1.00000

14 SECTIONS
.01 DB RIPPLE
G 0 1.00000
G 1 .83063
G 2 1.45735
G 3 1.84835
G 4 1.76443
G 5 1.98457
G 6 1.82135
G 7 2.01326
G 8 1.82899
G 9 2.00486
G10 1.80292
G11 1.94220
G12 1.67918
G13 1.60419
G14 .75462
G15 1.10075

15 SECTIONS
.01 DB RIPPLE
G 0 1.00000
G 1 .83212
G 2 1.46006
G 3 1.85206
G 4 1.76839
G 5 1.98978
G 6 1.82721
G 7 2.02161
G 8 1.83938
G 9 2.02161
G10 1.82721
G11 1.98978
G12 1.76839
G13 1.85206
G14 1.46006
G15 .83214
G16 1.00000

```
 2 SECTIONS        6 SECTIONS        10 SECTIONS       13 SECTIONS
 •10 DB RIPPLE     •10 DB RIPPLE     •10 DB RIPPLE     •10 DB RIPPLE
G 0  1•00000     G 0  1•00000     G 0  1•00000     G 0  1•00000
G 1   •84307     G 1  1•16813     G 1  1•19995     G 1  1•20740
G 2   •62202     G 2  1•40397     G 2  1•44819     G 2  1•45776
G 3  1•35538     G 3  2•05623     G 3  2•14445     G 3  2•16053
                 G 4  1•51709     G 4  1•62658     G 4  1•64142
                 G 5  1•90291     G 5  2•22535     G 5  2•25209
 3 SECTIONS       G 6   •86186     G 6  1•64186     G 6  1•67044
 •10 DB RIPPLE     G 7  1•35538     G 7  2•20464     G 7  2•26754
G 0  1•00000                      G 8  1•58218     G 8  1•67044
G 1  1•03158                      G 9  1•96286     G 9  2•25209
G 2  1•14740      7 SECTIONS       G10   •88533     G10  1•64142
G 3  1•03159     •10 DB RIPPLE     G11  1•35538     G11  2•16053
G 4  1•00000     G 0  1•00000                      G12  1•45776
                 G 1  1•18120                      G13  1•20742
                 G 2  1•42280                      G14  1•00000
 4 SECTIONS       G 3  2•09669     11 SECTIONS
 •10 DB RIPPLE     G 4  1•57339     •10 DB RIPPLE
G 0  1•00000     G 5  2•09669     G 0  1•00000      14 SECTIONS
G 1  1•10881     G 6  1•42280     G 1  1•20311     •10 DB RIPPLE
G 2  1•30618     G 7  1•18121     G 2  1•45229     G 0  1•00000
G 3  1•77038     G 8  1•00000     G 3  2•15146     G 1  1•20890
G 4   •81808                      G 4  1•63323     G 2  1•45964
G 5  1•35538                      G 5  2•23781     G 3  2•16357
                  8 SECTIONS       G 6  1•65591     G 4  1•64406
                 •10 DB RIPPLE     G 7  2•23781     G 5  2•25644
 5 SECTIONS       G 0  1•00000     G 8  1•63323     G 6  1•67455
 •10 DB RIPPLE     G 1  1•18978     G 9  2•15146     G 7  2•27508
G 0  1•00000     G 2  1•43465     G10  1•45229     G 8  1•67855
G 1  1•14684     G 3  2•11992     G11  1•20313     G 9  2•26965
G 2  1•37121     G 4  1•60100     G12  1•00000     G10  1•66480
G 3  1•97503     G 5  2•16997                      G11  2•22833
G 4  1•37121     G 6  1•56408                      G12  1•59628
G 5  1•14684     G 7  1•94450      12 SECTIONS       G13  1•97838
G 6  1•00000     G 8   •87782     •10 DB RIPPLE     G14   •89194
                 G 9  1•35538     G 0  1•00000     G15  1•35538
                                  G 1  1•20552
                                  G 2  1•45537
                  9 SECTIONS       G 3  2•15662
                 •10 DB RIPPLE     G 4  1•63794      15 SECTIONS
                 G 0  1•00000     G 5  2•24617     •10 DB RIPPLE
                 G 1  1•19570     G 6  1•66463     G 0  1•00000
                 G 2  1•44260     G 7  2•25620     G 1  1•21010
                 G 3  2•13457     G 8  1•65723     G 2  1•46116
                 G 4  1•61671     G 9  2•22004     G 3  2•16599
                 G 5  2•20539     G10  1•59115     G 4  1•64612
                 G 6  1•61671     G11  1•97259     G 5  2•25975
                 G 7  2•13457     G12   •88945     G 6  1•67756
                 G 8  1•44260     G13  1•35538     G 7  2•28037
                 G 9  1•19571                      G 8  1•68391
                 G10  1•00000                      G 9  2•28037
                                                   G10  1•67756
                                                   G11  2•25975
                                                   G12  1•64612
                                                   G13  2•16599
                                                   G14  1•46116
                                                   G15  1•21012
                                                   G16  1•00000
```

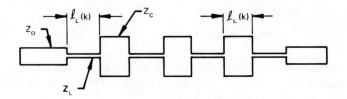

FIG. 6–11 Printed Realization of the High Impedance/Low-Impedance
Type of Low-Pass Filter

$$\ell_L(K) = \frac{V}{\omega} \text{ arc tan } \left[\frac{(G(K) \quad Z_0)}{Z_L} \right] \tag{6–27}$$

The length of the capacitance section, ℓ_c may be calculated by equation 6–28

$$\ell_c(K) = X_{cc}(K) \; Z_c \; \frac{V}{\omega} \tag{6–28}$$

$$X_{cc}(K) = \frac{G(K)}{Z_0} - \left[C_\pi(K-1) + C_\pi(K+1) \right] \tag{6–29}$$

$$C_\pi(K) = \frac{\omega \ell_K}{2 \, V \, Z_K} \tag{6–30}$$

C_π represents the unwanted capacitive susceptance of the high-impedance line which must be subtracted from the required capacity of the low-impedance section. In addition, the fringing capacity from the edge of the low-impedance line must also be subtracted from its total length in order to provide the proper value of shunt capacity. This can be calculated from equations 6–31 and 6–32. where C'$_f$ can be calculated from equation 2–5 and w equals the actual strip width in inches.

$$C = \frac{84.73 \quad \sqrt{\epsilon}}{Z_c} \; Pf/inch \tag{6–31}$$

$$\triangle \, \ell = \frac{0.450 \text{ W } \epsilon}{C} \left(\frac{C'_f}{\epsilon} \right) \tag{6--32}$$

The physical realization of this type of filter in stripline is not as good as that of the constant-K, m-derived, end-section filter. This is illustrated by a comparison of the responses in Figure 6–12.

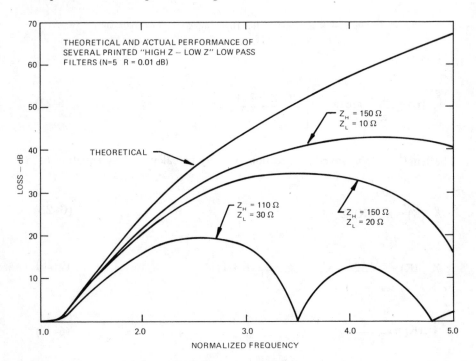

FIG. 6–12   Theoretical and Actual Performance of Several
High-Impedance/Low-Impedance,
Low-Pass Filters

It is imperative that the ratio between the high-impedance and low-impedance lines be kept as high as possible, consistent with both the frequency at which the filter is to be constructed and the minimum linewidths achievable within reasonable manufacturing tolerances.

Even with its degraded performance, there are many cases where this type of filter is a desirable device due to its small space requirement. However, additional sections beyond those which might be expected in theory will be necessary in order to achieve the required stop-band attenuation.

An additional type of low-pass filter which lends itself to printed stripline construction is the form shown in Figure 6–13. This circuit, which has both inductive and capacitive shunt elements, has a frequency response shown in Figure 6–14. The pass-band response has some amplitude ripple. The stop-band response has a series of peaks corresponding to the number of sections, and a minimum attenuation level, rather than the monotonically increasing attenuation value provided by Butterworth and Tchebyshev filters. Unlike the simple method of calculating the G-values for the conventional filter, the normalized L- and C-values of this class of filter have been derived by synthesis and are available from several sources in extensive tables.[3][4][5] Selection of some of the more useful cases for VSWR's of 1.02, 1.06, 1.1, 1.22, and 1.5, for filters having n = 3, 5, 7, and 9 sections have been compiled in Tables 6–4 through 6–7. In each case, the minimum attenuation and the minimum frequency at which this attenuation occurs are given. The values may be translated into inductance and capacitance by use of equations 6–25 and 6–26, or line dimensions for the configuration shown in Figure 6–15 by means of equations 6–27 through 6–32.

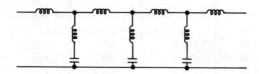

FIG. 6-13 General Schematic for Lumped-Element,
Elliptic-Function Low-Pass Filters

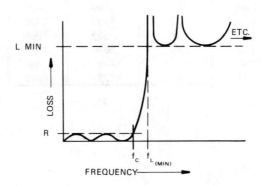

FIG. 6–14 Typical Response Curve for Elliptic-Function
Low-Pass Filters

TABLE 6 – 4
LOW PASS FILTER ELEMENTS

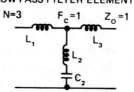

N=3 F_c =1 Z_o =1

VSWR	f_L	L_{MIN} dB	$L_1 = L_3$	L_2	C_2
1.02	2.79	10.4	.2490	.2227	.4399
1.02	4.13	20.7	.3050	.0803	.5509
1.06	2.00	10.3	.3646	.3456	.5615
1.06	2.92	20.8	.4518	.1222	.7290
1.06	4.13	30.2	.4891	.0552	.8007
1.10	2.00	14.5	.4915	.2760	.7031
1.10	2.46	20.4	.5438	.1585	.8004
1.10	3.63	31.1	.5968	.0640	.8991
1.22	1.47	10.7	.5776	.5885	.6362
1.22	2.00	20.4	.7160	.2230	.8701
1.22	2.92	31.3	.7920	.0892	.9992
1.22	4.13	40.7	.8233	.0420	1.0525
1.50	1.24	10.0	.7539	.9958	.5512
1.50	1.62	20.0	.9689	.3549	.8466
1.50	2.28	30.4	1.0855	.1466	1.0090
1.50	3.24	40.2	1.1395	.0669	1.0844

TABLE 6 – 5
LOW PASS FILTER ELEMENTS
N=5 $F_C=1$ $Z_O=1$

VSWR	f_L	L_{MIN} dB	L_1	L_2	C_2	L_3	L_4	C_4	L_5
1.02	1.56	21.9	.3566	.1910	.8809	1.0426	.8135	.4701	.0396
1.02	2.0	35.3	.4159	.0985	.9605	1.0979	.3201	.7156	.2555
1.02	2.92	53.4	.4559	.0416	1.0106	1.1629	.1182	.8999	.3886
1.06	1.31	20.5	.4709	.2716	.9699	1.1365	1.1820	.4654	.0715
1.06	1.56	31.4	.5428	.1560	1.0782	1.2211	.5257	.7274	.3055
1.06	2.0	44.8	.5942	.0822	1.1502	1.3155	.2422	.9457	.4659
1.06	2.92	63.0	.6290	.0351	1.1985	1.3934	.0962	1.1061	.5731
1.10	1.22	20.0	.5460	.3242	.9858	1.1656	1.4187	.4467	.1073
1.10	1.41	30.2	.6265	.1920	1.1099	1.2706	.6535	.7125	.3494
1.10	2.0	49.2	.7072	.0772	1.2257	1.4348	.2206	1.0383	.5884
1.10	2.92	67.4	.7406	.0330	1.2732	1.5182	.0895	1.1884	.6883
1.22	1.15	21.4	.7083	.3837	.9948	1.2204	1.5864	.4501	.2483
1.22	1.31	31.0	.8003	.2332	1.1301	1.3675	.7740	.7108	.4888
1.22	1.74	48.2	.8932	.1022	1.2603	1.5768	.2914	1.0395	.7418
1.22	2.00	55.3	.9151	.0733	1.2909	1.6348	.2034	1.1264	.8039
1.22	2.92	73.5	.9479	.0315	1.3366	1.7270	.0843	1.2621	.8983
1.5	1.15	27.5	1.0247	.3789	1.0076	1.4125	1.3297	.5370	.5944
1.5	1.31	37.1	1.1203	.2342	1.1251	1.6134	.7148	.7697	.8177
1.5	1.74	54.3	1.2176	.1036	1.2429	1.8677	.2867	1.0565	1.0661
1.5	2.00	61.4	1.2407	.07446	1.2709	1.9355	.2024	1.1320	1.1287
1.5	2.92	79.6	1.2751	.0321	1.3129	2.0418	.0852	1.2498	1.2249

TABLE 6 – 6
LOW PASS FILTER ELEMENTS
$N=7$ $F_c=1$ $Z_0=1$

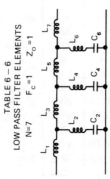

VSWR	f_L	L_{MIN} dB	L_1	L_2	C_2	L_3	L_4	C_4	L_5	L_6	C_6	L_7
1.02	1.31	36.1	.4249	.1489	1.0331	1.0440	.7284	.7801	.9720	.6776	.6157	.1016
1.02	1.56	51.4	.4691	.0876	1.0914	1.1938	.3849	1.0325	1.1235	.3404	.8167	.2773
1.02	2.00	70.1	.4996	.0466	1.1317	1.3112	.1939	1.2336	1.2633	.1654	.9750	.3966
1.06	1.22	38.8	.5742	.1642	1.1570	1.1405	.8265	.7891	1.0193	.6938	.7103	.2470
1.06	1.56	61.0	.6394	.0764	1.2510	1.3792	.3419	1.1625	1.2934	.2762	1.0067	.4776
1.06	2.00	79.7	.6669	.0409	1.2907	1.4979	.1767	1.3538	1.4451	.1400	1.1518	.5780
1.10	1.15	36.6	.6556	.1972	1.1747	1.1371	1.0475	.6988	.9854	.8389	.6805	.2864
1.10	1.31	50.1	.7107	.1226	1.2552	1.3239	.5856	.9699	1.1951	.4628	.9015	.4657
1.10	1.56	65.4	.7488	.0729	1.3110	1.4780	.3310	1.2009	1.3860	.2572	1.0810	.5962
1.10	2.00	84.1	.7755	.0391	1.3502	1.5992	.1725	1.3868	1.5440	.1322	1.2196	.6910
1.22	1.15	42.7	.8602	.1892	1.224	1.2869	.9889	.7402	1.0966	.7384	.7731	.5188
1.22	1.31	56.2	.9143	.1183	1.3006	1.4840	.5711	.9949	1.3357	.4259	.9796	.6829
1.22	1.56	71.5	.9520	.0706	1.3542	1.6465	.3278	1.2126	1.5449	.2426	1.1458	.8060
1.22	2.00	90.2	.9785	.0379	1.3920	1.7744	.1724	1.3879	1.7148	.1266	1.2737	.8971
1.50	1.15	48.8	1.1768	.1939	1.1947	1.5113	1.0110	.7240	1.2778	.7121	.8017	.8360
1.50	1.31	62.6	1.2336	.1218	1.2638	1.7309	.5966	.9525	1.5567	.4235	.9850	.9992
1.50	1.56	77.6	1.2735	.0728	1.3127	1.9118	.3462	1.1480	1.7952	.2456	1.1319	1.243
1.50	2.00	96.3	1.3017	.0391	1.3473	2.0541	.1833	1.3055	1.9867	.1295	1.2448	1.2179

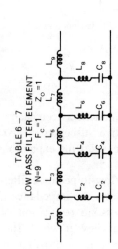

TABLE 6 – 7
LOW PASS FILTER ELEMENT
N=9 $F_c=1$ $Z_o=1$

VSWR	f_L	L_{MIN} dB	L_1	L_2	C_2	L_3	L_4	C_4	L_5	L_6	C_6	L_7	L_8	C_8	L_9
1.10	1.03	31.1	.6202	.2616	1.127	.9578	1.884	.4664	.5256	3.114	.3005	.6835	1.338	.5181	.1245
1.10	1.06	41.4	.6737	.1890	1.206	1.107	1.210	.6594	.7037	1.804	.4849	.8338	.8531	.6911	.2883
1.10	1.15	57.9	.7292	.1176	1.288	1.308	.6882	.9381	.9958	.9487	.7791	1.0750	.4771	.9154	.4695
1.10	1.31	75.3	.7643	.0736	1.341	1.467	.4130	1.173	1.261	.5504	1.0460	1.290	.2821	1.085	.5932
1.10	1.56	95.0	.7883	.0440	1.377	1.593	.2413	1.365	1.488	.3163	1.276	1.473	.1627	1.215	.6824
1.22	1.03	37.1	.8242	.2509	1.175	1.090	1.771	.4961	.5783	2.794	.3350	.7555	1.135	.6106	.3721
1.22	1.06	47.4	.8761	.1826	1.245	1.245	1.176	.6786	.7784	1.710	.5116	.9308	.7617	.7741	.5162
1.22	1.15	64.0	.9306	.1143	1.325	1.455	.6846	.9430	1.095	.9336	.7917	1.198	.4439	.9837	.6832
1.22	1.31	81.4	.9655	.0718	1.375	1.621	.4156	1.165	1.378	.5508	1.045	1.431	.2682	1.142	.8005
1.22	1.56	101.0	.9894	.0429	1.410	1.754	.2443	1.348	1.618	.3195	1.264	1.626	.1567	1.262	.8867
1.50	1.015	35.8	1.084	.3292	1.082	1.146	2.538	.3645	.5094	1.276	.2291	.7286	1.455	.5281	.5658
1.50	1.03	43.3	1.137	.2572	1.146	1.283	1.807	.4862	.6681	2.768	.3381	.8782	1.068	.6491	.6902
1.50	1.06	53.6	1.191	.1882	1.211	1.455	1.227	.6507	.8992	1.754	.4987	1.085	.7413	.7953	.8294
1.50	1.15	70.1	1.248	.1183	1.281	1.686	.7261	.8891	1.257	.9825	.7523	1.392	.4449	.9816	.9959
1.50	1.31	87.5	1.285	.0745	1.326	1.871	.4444	1.090	1.574	.5867	.9811	1.655	.2730	1.1210	1.116
1.50	1.56	107.2	1.310	.0446	1.357	2.017	.2626	1.254	1.842	.3426	1.178	1.874	.1610	1.228	1.204

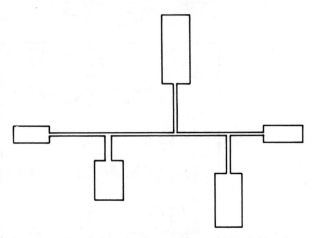

FIG. 6–15   Physical Realization of a Printed Elliptic-Function,
Low-Pass Filter

Inasmuch as there are three inductive lines, rather than one, between each of the capacitive sections, the additional unwanted capacitive susceptance given in equation 6–30 must be calculated for each of these cases, thus expanding the negative term in equation 6–29.

As in the case of the previous high-impedance/low-impedance filter it is desirable to keep the line lengths as short as possible and the impedance ratios as high as possible in order to provide the best possible realization of the theoretical response, as well as to maximize the frequency at which the first spurious response occurs. The one exception to this rule is the first shunt-inductive section, which typically becomes very short unless a lower impedance is used.

As a result of the ripple stop-band and because the element lengths of these filters become a significant fraction of a wavelength, the upper frequency limit of the stop-band characteristic may not be sufficient. Therefore, it may be necessary to add a conventional low-pass filter set at some higher frequency of cutoff in order to provide both sharp skirts and sustained stop-band attenuation. Typical performance for a printed filter versus a lumped constant filter is shown in Figure 6–16. It illustrates once again the fact that printed versions of lumped-constant filters do not provide the theoretical performance which may be expected of them.

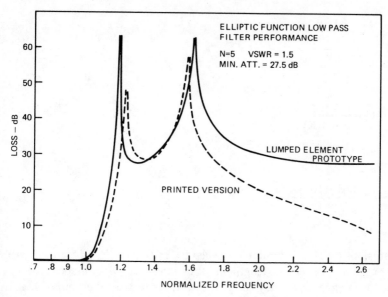

FIG. 6–16   Theoretical vs Actual Response for Printed
Elliptic-Function, Low-Pass Filters

High-Pass Filters

High-pass filters may be conveniently constructed by inverting the G-values as shown in Figure 6–7b. In this case, the values of inductance become capacitance, and vice versa. The response curves given in Figures 6–8, 6–9, and 6–10 may also be reversed to show the roll-off on the low side as opposed to the roll-off on the high side. Again, as in the case of the low-pass structure, an uneven number of sections is necessary to construct filters when using the design equations given here. The capacitive sections will be calculated by using equation 6–33, which has been denormalized to provide actual capacitance values expressed in picofarads. This is because it is frequently desirable in the case of the high-pass filter to use lumped-element chip-type capacitors where the element values can be achieved by this method. This is particularly true at the lower frequencies where the values of capacity may be high, resulting in capacitive structures with excessively large dimensions unless some lumped-element device is used. The shunt inductances may be calculated from equation 6–34 and can then be translated into line lengths of a given impedance, Z_1, by means of equation 6–35.

$$C(K) = \frac{\left(\dfrac{1}{2\pi F \ G(K)}\right)}{Z_o} \left(10^{12}\right) \ Pf \tag{6-33}$$

$$L(K) = \left(\frac{1}{2\pi \ F \ G(K)}\right) Z_o \tag{6-34}$$

Where F is in MHz

$$\ell_\ell(K) = \frac{L(K) \ (10^9) \ (\sqrt{\epsilon})}{(.0847 \ Z_\ell)} \tag{6-35}$$

As in previous cases, it is desirable to keep the impedance Z_1 as high as possible in order to minimize the line length and thus extend the performance of the pass band to as high a frequency as possible. In this case, the shunt-capacitive susceptance of the inductive line cannot be readily compensated, as it can be in the case of filters having shunt-capacitive sections, and thus shows up as an error in the construction of the printed version of the filter. It is, however, a small error and does not appear to have as great a significance in the filter as the errors which occur in the low pass versions. This can be seen in Figure 6–17, which shows the theoretical and practical printed version response of two high-pass filters. While the deviations from theoretical response are not as violent as those in the low-pass case, the insertion loss of the pass-band increases rapidly at frequencies greater than twice the frequency of cutoff. Although major spurious responses will not occur until even higher frequencies, this gradual increase in insertion loss may be intolerable. A possible configuration for construction of the high-pass filter in printed circuit line is shown in Figure 6–18. This approach makes use of three-layer construction permitting overlaps to generate the capacitive coupling sections. The degree of overlap may be calculated by making use of the standard parallel plate capacity equations and is given in equation 6–36.

$$O(K) = \frac{T \ (C(K))}{(.225 \ \epsilon \ W)} \tag{6-36}$$

Where   T = Spacer thickness   (in inches)
        W = Strip width

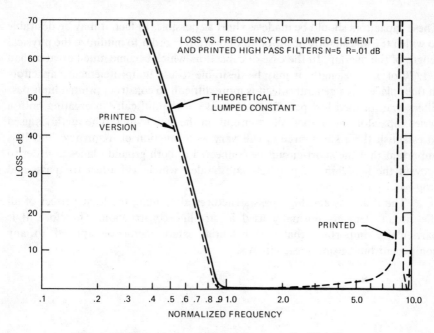

FIG. 6–17   Theoretical and Printed Responses for High-Pass
Filters

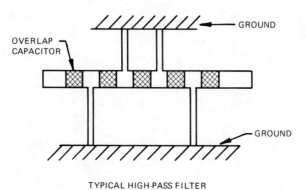

TYPICAL HIGH-PASS FILTER

FIG. 6–18

These capacitors should be made as short as possible. In fact, it may be desirable to widen the line slightly at the point of overlap in order to minimize the physical length of the overlap. In the case of capacitors which become much greater than a 1/20 of a wavelength, it may be desirable to use lumped-element capacitors in their place. As a general rule, it is more difficult to construct printed high-pass filters than printed low-pass filters, because of the difficulty in creating both a series capacitor and a good short-circuit on the shunt stubs. The methods used to establish these short-circuits will vary as a function of frequency, but, it is important that the short-circuit be connected to both ground planes in order to prevent the launching of parallel-plate modes, which will affect the pass- and stop-band performances.

The low-pass and high-pass structures are among the least perfect of all the circuit elements normally used in stripline construction. Therefore, it is particularly important that a substantial safety factor be applied to any component built using these circuits.

REFERENCES

1 Peters, B. W., et al, *Handbook of Tri-Plate Microwave Components,* Sanders Associates, Nashua, New Hampshire, 1956, pp. 89–102.

2 Matthei, G., Jones, E. M. T., Young, L., *Microwave Filters Impedance Matching Networks and Coupling Structures,* McGraw Hill, New York, 1964, pp. 85–94.

3 Saal, R., and Ulbrich, E., *On The Design of Filters by Synthesis,* IRE Transactions on Circuit Theory, December 1958, pp. 284–327.

4 Saal, R., *Der Entwurf von Filtern mit Hilfe des Kataloges Normierter Tiefpasse,* Telefunken GMBH, Backnang, W. Germany, 1963.

5 Zverev, A. I., *Handbook of Filter Synthesis,* John Wiley and Sons, New York, 1967.

6 Matthei, Jones and Young, *Microwave Filters Impedance Matching Networks and Coupling Structures,* McGraw Hill, New York, 1964, pp. 355–380.

CHAPTER

7

Bandpass and Bandstop Filters

Introduction

There are multitudes of methods of constructing microwave bandpass and bandstop filters, many of which have direct application to stripline circuit construction. Therefore, the emphasis of Chapter 7 will be those techniques which can be readily employed by the stripline circuit designer.

Virtually all of these filters base their responses on low-pass prototypes and thus, roll-off, skirt, and other filter characteristics can be determined from examining the low-pass filter response curves in Figures 6–8, 6–9, and 6–10. Those figures express filter skirt roll-off in terms of normalized cut-off frequency $F = 1$ represents the cutoff frequency of the low-pass filter, $F = 2$ occurs at a frequency of twice the cutoff frequency, etc. For the bandpass case, the normalized frequency of the curves can be related to half-bandwidths of the bandpass response. Thus, for example, if a bandpass filter has a bandwidth of 200 MHz, and is centered at 1.0 GHz, $F = 1$ will represent the frequency response at the upper edge of the band, or 1.1 GHz. The response of $F = 2$ will be equal to that at a frequency two half-bandwidths from the center frequency of the filter or, in this example 1.2 GHz. As in the case of the high-pass filter, the curves can be inverted to give the response for the lower edge of the filter as well. The low-pass prototype is useful, not only in determination of the skirt response of the filter, but also for the actual design of circuits to be built. Thus, the G-values established in Tables 6–1, 6–2, and 6–3, will be used in the design of bandpass filters in this chapter.

TEXAS STATE TECHNICAL INSTITUTE
ROLLING PLAINS CAMPUS – LIBRARY
SWEETWATER, TEXAS 79556

Insertion Loss

The passband insertion loss of any bandpass filter is a function of the filter's relative bandwidth and the available unloaded Q of the transmission line being used to fabricate it. In general the insertion loss of the filter increases as its bandwidth decreases. It will also increase as the number of resonators in the filter increases, although not always at the same rate. Thus, it may be desirable, where sharp skirts and narrow bandwidths are required, to broaden the bandwidth slightly and to increase the number of sections in order to maintain the necessary stop band rejection while minimizing the insertion loss of the passband.

For any bandwidth or any number of resonators, the fundamental governing limit on insertion loss is the available unloaded Q where

$$\frac{1}{Q_u} = \frac{1}{Q_c} + \frac{1}{Q_d} \qquad\qquad (7\text{-}1)$$

Q_c and Q_d represent the unloaded Q contribution as a result of copper losses and dielectric losses. Q_c may be calculated from equation 7–2 where A_c is as defined in equations 1–4 or 1–5.

$$Q_c = \frac{2\pi \sqrt{\epsilon}}{\lambda\, a_c} \qquad\qquad (7\text{–}2)$$

$$Q_d = \frac{1}{Tan\ \delta} \qquad\qquad (7\text{-}3)$$

Q_d shown in equation 7–3 is related only to the loss tangent of the dielectric which may be found in the tables of Chapter One. Unloaded Q versus frequency for 50 ohm lines has been calculated for the major dielectric mediums discussed. Significantly, the major source of loss, and thus the limit to unloaded Q, is the dielectric, rather than the copper, losses. Therefore, there is only a very small change (less than 10%) between the unloaded Q for a 1/16" ground-plane spacing and 1/8" ground-plane spacing, or between a 1/8" ground-plane spacing and a 1/4" ground-plane spacing. Since the values of unloaded Q cannot be predicted accurately due to construction limitations, gaps, dielectric quality, and other factors, the values of unloaded Q which have been calculated for the 1/8" case are adequate for most calculations. These are shown in Figure 7–1.

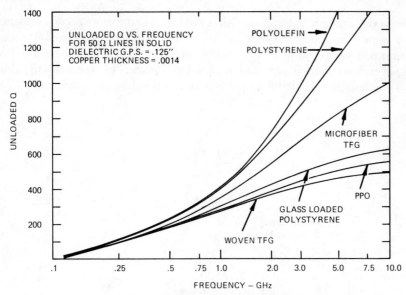

FIG. 7–1   Unloaded Q vs. Frequency for 50 Ohm
Line in Solid Dielectric Stripline

For a given bandpass filter, the insertion loss will be the sum of the losses of each
of the resonators as shown in equation 7–4.

$$Loss_{TOT.} \ (dB) = \sum_{R=1}^{n} L_R \tag{7-4}$$

The individual resonator losses are given by equation 7–5

$$L_R = 20 \ Log_{10} \left[\frac{Q_u}{Q_u - Q_L} \right] \tag{7-5}$$

where Q_u is the unloaded Q of each resonator and Q_L is the loaded Q of each
resonator. Q_u will vary from resonator to resonator as a function of its coupling
and dimensions. For general approximations, the unloaded Q of the 50 ohm line
is sufficient and may be taken from the curve of Figure 7–1. As a practical matter,

however, many workers in the field have found that a more accurate prediction of actual insertion loss can be made if a value of unloaded Q equal to approximately 60% to 75% of the theoretical unloaded Q is used. The loaded Q of each resonator is expressed by equations 7–6 and 7–7, where R is the location of the resonator in the filter and n is the total number of resonators.

$$Q_L = Q_{L_{TOT.}} \, \text{Sin} \, \left(\frac{2\Omega - 1}{2n} \right) \, \pi \qquad (7\text{-}6)$$

$$Q_{L_{TOT.}} = \frac{fo}{f2 - f1} \qquad (7\text{-}7)$$

where Ω = position of the resonator
and n = number of resonators

The insertion loss as calculated by this technique will be reasonably accurate; however, additional losses for connectors, interfaces, mismatches and the inaccuracies of the values of unloaded Q should be anticipated, particularly in the case of extremely low-loss filters.

Half-Wave Resonator, Side-Coupled Filter

The most common bandpass filter in general use for stripline applications is the side-coupled filter making use of open-circuit, half-wave resonators. This same basic filter may be constructed with quarter-wave short-circuited resonators, but for stripline applications that technique raises the problem of generating a good quality short-circuit. A half-wave resonator, on the other hand, can be readily printed and has the additional advantage of providing dc isolation. Its general configuration is shown in Figure 7–2.

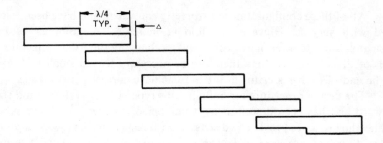

SIDE COUPLED HALF-WAVE RESONATOR
BANDPASS FILTER

FIG. 7–2   Side-coupled Half-wave Resonator Stripline Filter
Configuration

The use of half-wave resonators has the disadvantage of making a filter larger than those employing quarter-wave resonators. Thus, in most cases, it becomes long and thin, and may not be convenient to package in an overall stripline circuit. As a result, a number of methods have been developed for folding, or otherwise configuring, side-coupled filters. These are shown in Figure 7–3, which describes a pyramid type of fold, a hairpin type of construction, and a pseudo-interdigital type of filter.

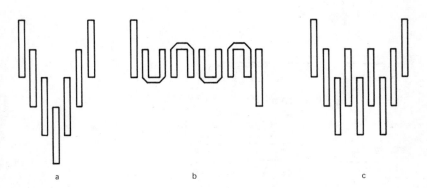

a   b   c

FIG. 7–3   Common Methods of Folding Side-Coupled Half-Wave
Filters

All of these configurations are directly equivalent and have been tested and used with success. However, as folding increases, the number of internal mismatches in the filter increases as well as changing the fringe capacities of the ends of the sections. It may therefore be necessary to add small tuning screws at the end of each resonator in order to achieve correct performance.

The design procedure for building this type of filter is simple and straight-forward and has been described by a number of sources.[1][2][3][4] The procedure relates the G-values previously discussed with respect to the low-pass prototype. These G-values are applied in the calculation of the normalized even-mode impedances which may then be used by means of the equations and curves in Chapter 4 to develop the dimensions of the individual sections, and the gaps between the sections. The normalized even-mode impedances are given by equation 7–8.

$$Z_{oe(norm)} = 1 + \frac{J_{k,k+1}}{Yo} + \left(\frac{J_{k,k+1}}{Yo}\right)^2$$

$$k, k+1 \Big|_{k=0 \text{ to } n}$$

$$\tag{7-8}$$

$$\frac{Jo_1}{Yo} = \sqrt{\frac{\pi \, \omega}{2 \, g_o \, g_1}} \tag{7-9}$$

$$\frac{J_{k,k+1}}{Yo} \Big|_k = \frac{\pi \, \omega}{2} \frac{1}{\sqrt{g_k \, g_{k+1}}} \tag{7-10}$$

$$k = 1 \text{ to } n-1$$

$$\frac{J_{n,n+1}}{Yo} = \sqrt{\frac{\pi \, \omega}{2 \, g_n \, g_{n+1}}} \tag{7-11}$$

$$\omega = \frac{B.W.}{fo} = \frac{f2 - f1}{fo} \tag{7-12}$$

Nominally, each coupled section will be one quarter-wavelength long at the center frequency of the filter. In actual practice, this quarter-wavelength must be foreshortened slightly in order to compensate for the end fringe-capacity of the resonator. This foreshortening will vary slightly from resonator to resonator based on the coupling and the linewidth of each section.[5] For most filters, the amount of foreshortening may be calculated from equation 7-13[6] where $\triangle$ equals the amount of foreshortening at each end of the resonator, and b is the total ground-plane spacing.

$$\triangle = .165 \text{ (b)} \tag{7-13}$$

As a convenience, normalized even-mode impedances for each of the coupling sections for filters having maximally flat responses and equal-ripple responses are given in Figures 7–4 through 7–16.

All the half-wave resonator filters presented have second-order responses at three times the fundamental frequency. This second response will have approximately the same bandwidth as the primary response. In the case of a filter which has been improperly constructed, a spurious response may also occur at twice the fundamental frequency. This condition occurs only to half-wave resonator filters and does not apply to the quarter-wave filters described later.

End-Coupled Resonator Filters

An alternate type of half-wave resonator filter is the end-coupled filter shown in Figure 7–17. In this case the resonators are coupled by small capacitive gaps, thus creating a straight line version which is approximately twice as long as the half-wave side-coupled filter previously described. This filter can, of course, be meandered in a serpentine fashion in order to minimize its overall length, or, in higher frequency versions where this is not a problem, built in the straight line configuration shown. The susceptances for each section may be calculated from equation 7–14,[7] where the values of J_K/Y_0 are determined from equations 7–9, 7–10, and 7–11.

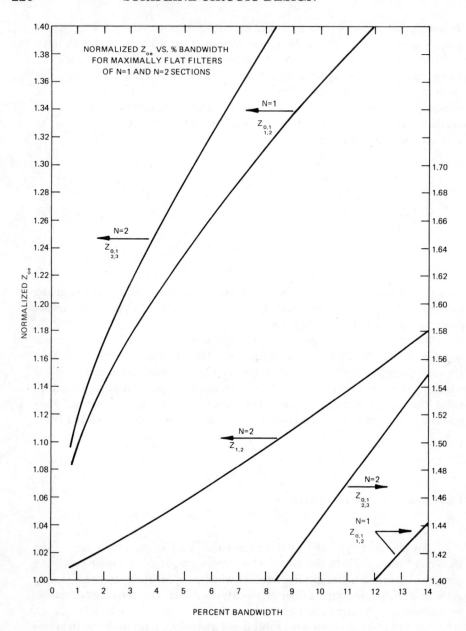

FIG. 7–4 Normalized Even-Mode Impedance vs. Percentage
Bandwidth for Maximally Flat Filters Having One or
Two Sections

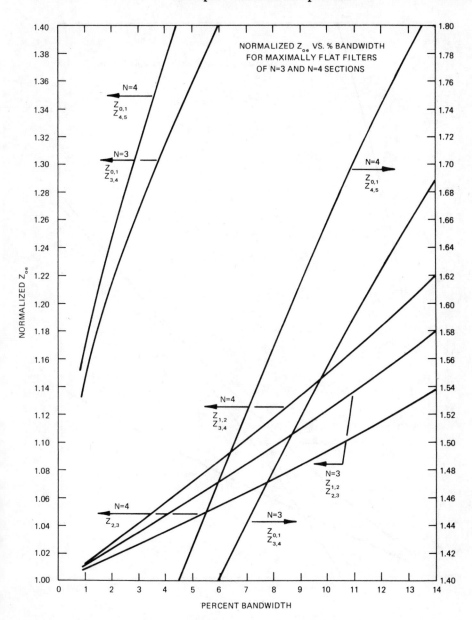

FIG. 7–5   Normalized Even-Mode Impedance vs. Percentage
Bandwidth for Maximally Flat Filters Having Three or
Four Sections

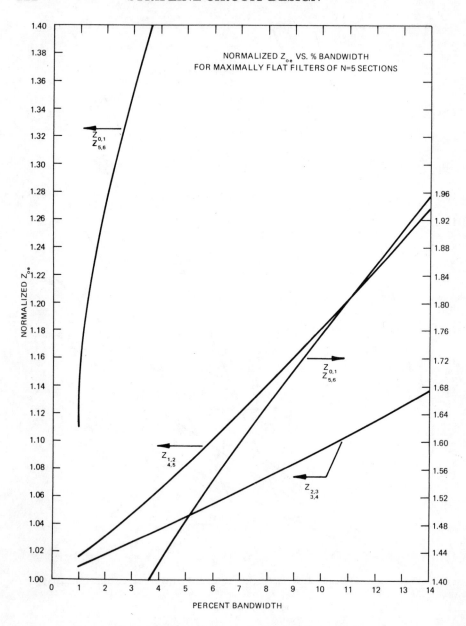

FIG. 7–6 Normalized Even-Mode Impedance vs. Percentage
Bandwidth for Maximally Flat Filters Having
Five Sections

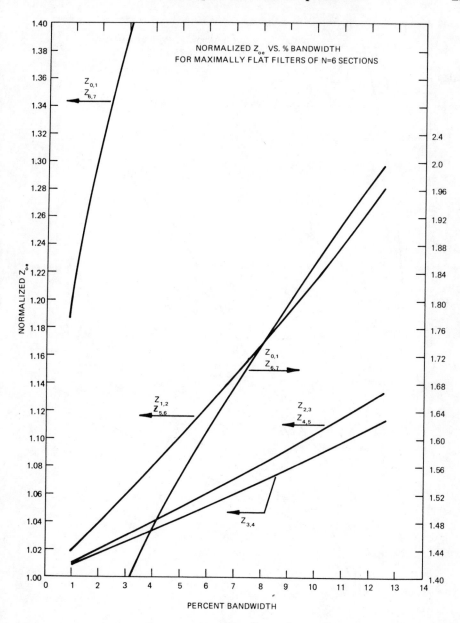

FIG. 7–7 Normalized Even-Mode Impedance vs. Percentage
Bandwidth for Maximally Flat Filters Having
Six Sections

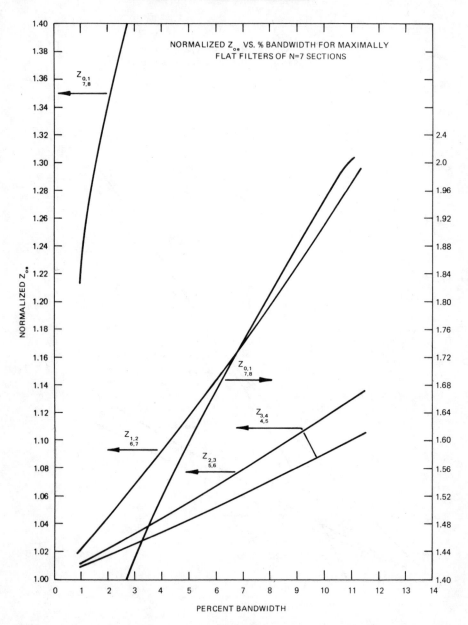

FIG. 7–8 Normalized Even-Mode Impedance vs. Percentage
Bandwidth for Maximally Flat Filters Having
Seven Sections

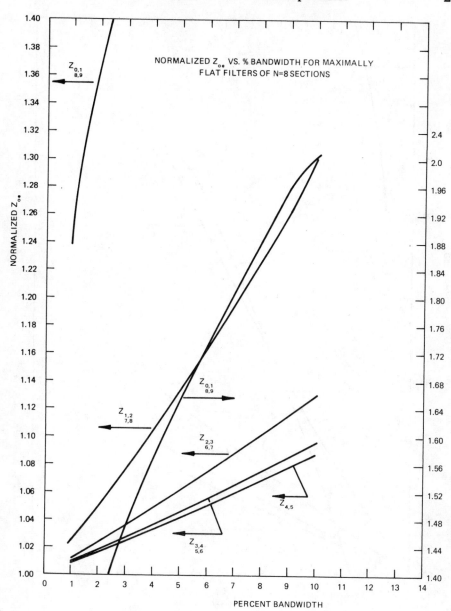

FIG. 7–9 Normalized Even-Mode Impedance vs. Percentage
Bandwidth for Maximally Flat Filters Having
Eight Sections

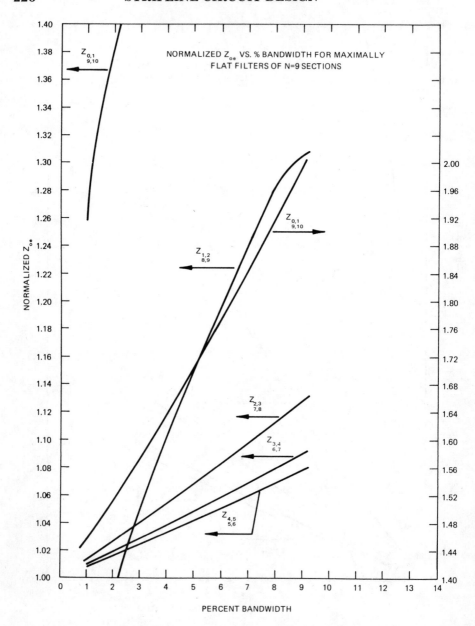

FIG. 7–10 Normalized Even-Mode Impedance vs. Percentage
Bandwidth for Maximally Flat Filters Having
Nine Sections

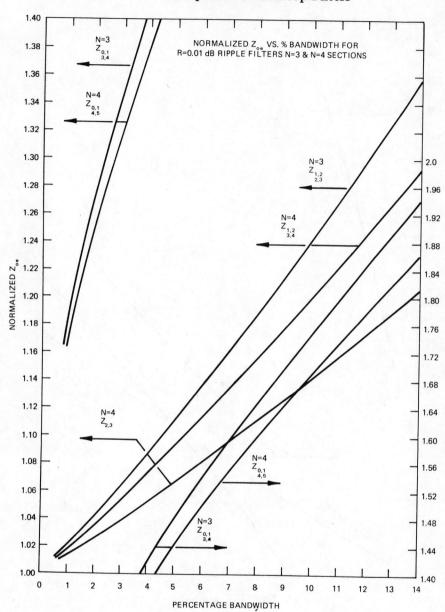

FIG. 7–11 Normalized Even-Mode Impedance vs. Percentage
Bandwidth for 0.01 dB Ripple Filters Having
Three and Four Sections

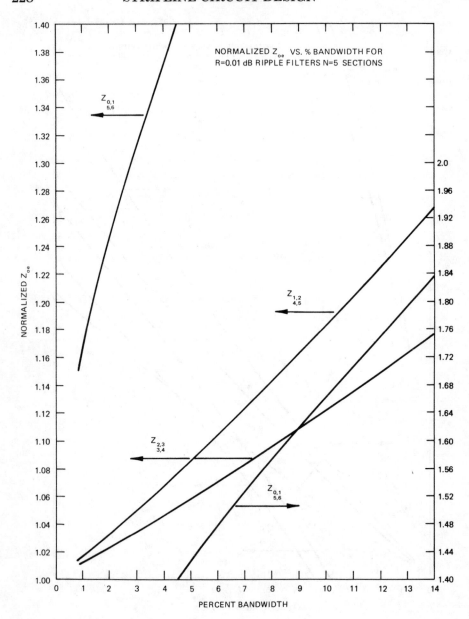

FIG. 7–12   Normalized Even-Mode Impedance vs. Percentage
for 0.01 dB Ripple Filters Having Five Sections

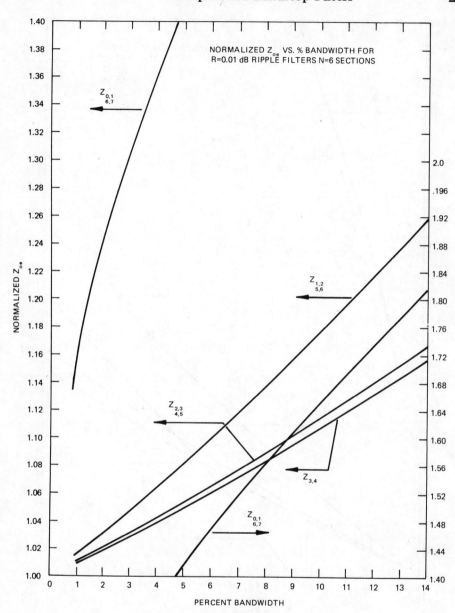

FIG. 7–13 Normalized Even-Mode Impedance vs. Percentage
Bandwidth for 0.01 dB Ripple Filters Having Six Sections

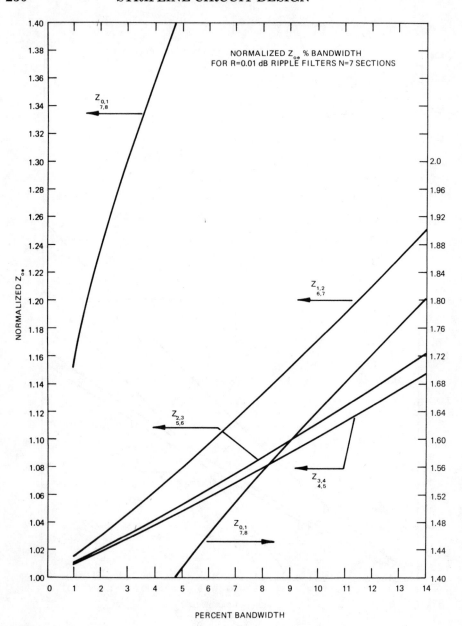

FIG. 7–14 Normalized Even-Mode Impedance vs. Percentage
Bandwidth for 0.01 dB Ripple Filters Having Seven Sections

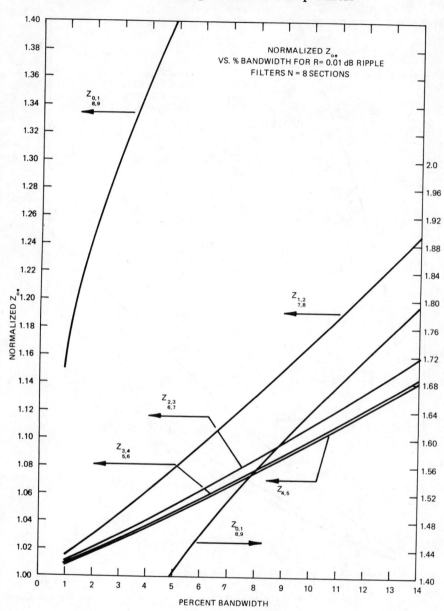

FIG. 7–15 Normalized Even-Mode Impedance vs. Percentage
Bandwidth for 0.01 dB Ripple Filters Having
Eight Sections

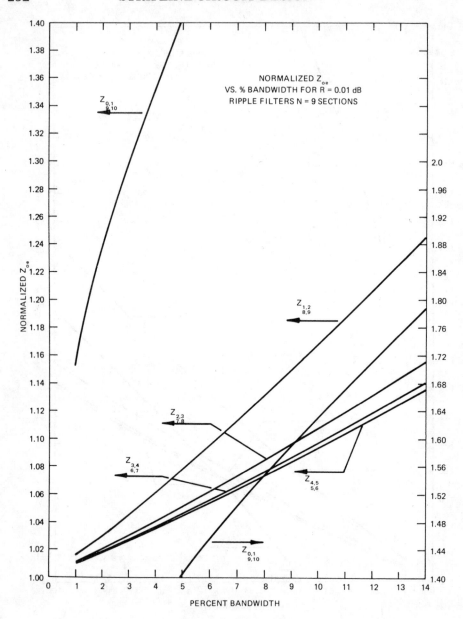

FIG. 7–16 Normalized Even-Mode Impedance vs. Percentage
Bandwidth for 0.01 dB Ripple Filters Having
Nine Sections

$$\frac{B_{k,\,k+1}}{Yo} = \frac{\sqrt{\dfrac{J_{k,\,k+1}}{Yo}}}{1 - \left(\dfrac{J_{k,\,k+1}}{Yo}\right)^2} \tag{7-14}$$

The dimensions of the individual gaps may be determined by using Oliner's gap-equivalent circuit shown in Figure 7–18[8][9] with the dimensions calculated from equations 7–15 and 7–16, where d is the gap, and b is the ground plane spacing.

$$\frac{B_a}{Yo} = -\frac{2b}{\lambda}\,\log_e\left[\cosh\left(\frac{\pi D}{b}\right)\right] \tag{7-15}$$

$$\frac{B_b}{Yo} = \frac{b}{\lambda}\,\log_e\left[\coth\left(\frac{\pi D}{2b}\right)\right] \tag{7-16}$$

These equations are quite accurate for zero strip-thickness cases where w/b $\geq$ 1.2. This condition exists for most 50 ohm line filters. However, if the impedance of the filter increases, with the resultant decrease in stripwidth, substantial errors in the gap dimensions can occur. The length of each resonant section from the center line of the gap to the center line of the next gap can be determined from equation 7–17 and 7–18.

$$\theta_{k_{\text{Radians}}} = \pi - 1/2\left[\tan^{-1}\left(\frac{2B_{k-1,\,k}}{Yo}\right) + \tan^{-1}\left(\frac{2B_{k,\,k+1}}{Yo}\right)\right] \tag{7-17}$$

$$\theta_{k}\Big|_{k=1\text{ to }n} = \pi + 1/2\left[\theta_{k-1,\,k} + \theta_{k,\,k+1}\right] \tag{7-18}$$

Because the gap dimensions for very wide band filters become so small that they are not physically achievable even with interdigitated gaps, the usefulness of this type of filter is restricted to extremely narrow bandwidths, typically less than 5%. As a result, it has limited usefulness and has not achieved great popularity,

 END-COUPLED RESONATOR
BAND-PASS FILTER

FIG. 7–17 Basic Configuration for End-Coupled Resonator Bandpass Filters

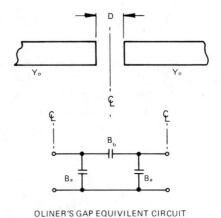

OLINER'S GAP EQUIVILENT CIRCUIT

FIG. 7–18 Oliner's Gap Equivalent Circuit for End-Coupled Sections

although short overlap sections as described for high pass filter construction in Chapter 6, could undoubtedly be used to reduce the sensitivity to manufacturing tolerances and increase the bandwidth of this device.

Direct-Coupled Stub Filters

Parallel– and end-coupled half-wave filters are generally most useful for narrow and medium bandwidth applications. For wideband applications of 30% or greater, the most effective type of filter with the least manufacturing tolerance sensitivity is a direct-coupled stub filter as shown in schematic form in Figure 7–19.

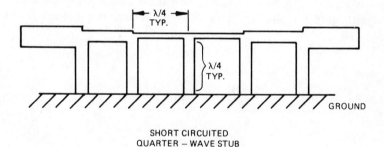

SHORT CIRCUITED
QUARTER – WAVE STUB
BAND PASS FILTER

FIG. 7–19   Basic Configuration for Short Circuited Quarter-Wave
Stub Bandpass Filters

This filter consists of a series of short-circuited quarter-wavelength stubs separated by quarter-wavelength sections of line. Each shorted stub represents one pole or section of the filter. Because it is frequently difficult to achieve the actual impedance levels necessary for single quarter-wave short-circuited sections, alternate sections may be used. For example, Figure 7–20 shows equivalent circuit elements which may be used in place of the single-section quarter-wave shorted stub. These include two quarter-wave shorted stubs at twice the impedance of each section, a single half-wave open-circuited stub at twice the impedance of each section, or even two open-circuited half-wave stubs, each one being four times the impedance of each section. These transformations, in addition to main line impedance transformation using the transformers described

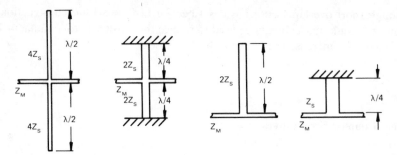

FIG. 7–20　Equivalent Stub Structures for
Bandpass Filters

in Chapter 3, may be used to generate filters having bandwidths from 20% to ratios as great as 3:1, while maintaining impedance levels within the normal range of construction for stripline devices. Although this type of filter is not new, a recently published paper by Cristal[10] provides a design technique which is both straightforward and effective. This technique is given in equations 7–19 through 7–31.

$$Y_1 = \frac{A_{22}(1) + A_{11}(2) - A_{12}(1) \left[A_{12}(1) + A_{12}(2) \right]}{(A_{12}(1))^2} \qquad (7\text{-}19)$$

$$Y_2 = A_{11}(2) + A_{11}(3) - \frac{A_{12}(2)}{A_{12}(1)} - A_{12}(3) \qquad (7\text{-}20)$$

$$Y_k = A_{11}(k) + A_{11}(k+1) - \left[A_{12}(k) + A_{12}(k+1) \right]$$
$$\text{for } k = 3, 4, \ldots\ldots, (n-2) \qquad (7\text{-}21)$$

$$Y_{(n-1)} = A_{11}(n-1) + A_{11}(n) - \frac{A_{12}(n)}{A_{12}(n+1)} - A_{12}(n-1) \qquad (7\text{-}22)$$

$$Y_n = \frac{A_{22}(n+1) + A_{11}(n) - A_{12}(n+1) \left[A_{12}(n+1) + A_{12}(n) \right]}{\left(A_{12}(n+1) \right)^2} \qquad (7\text{-}23)$$

$$Y_{12} = \frac{A_{12}(2)}{A_{12}(1)} \tag{7-24}$$

$$Y_{k, k+1} = A_{12}(k+1)$$
$$\text{for } k = 2, 3, \ldots \ldots, (n-2) \tag{7-25}$$

$$Y_{n-1, n} = \frac{A_{12}(n)}{A_{12}(n+1)} \tag{7-26}$$

For $k = 1$ and $N + 1$ For $k = 2, 3, \ldots \ldots, n$

$$A_{11}(k) = 1 \qquad\qquad\qquad A_{11}(k) = h\tau$$
$$A_{12}(k) = \sqrt{h \ G_{(k)}} \qquad\qquad A_{12}(k) = h \ G_{(k)} \sin \theta \tag{7-27}$$
$$A_{22}(k) = \left[h \ G_{(k)}^2 + \tau \right]$$

$$\theta_1 = \frac{\pi}{2} \left(1 - \frac{\omega}{2} \right) \tag{7-28}$$

$$\tau = \frac{1}{2} \tan \theta_1 \tag{7-29}$$

$$G_{(k)} = \frac{1}{\sqrt{g_{(k-1)} \ g_k}} \tag{7-30}$$

$h =$ an arbitrary positive dimensionless parameter normally
       less than or equal to one, which controls the internal $(7-31)$
       immitance level

As in the case of the side-coupled parallel-section filter, the usefulness of this circuit is such that design curves are being provided for maximally flat and 0.01 equal-ripple filters of 2 to 9 sections. These can be found in Figures 7–21 through 7–34 and are expressed in terms of the normalized impedance versus percentage bandwidth.

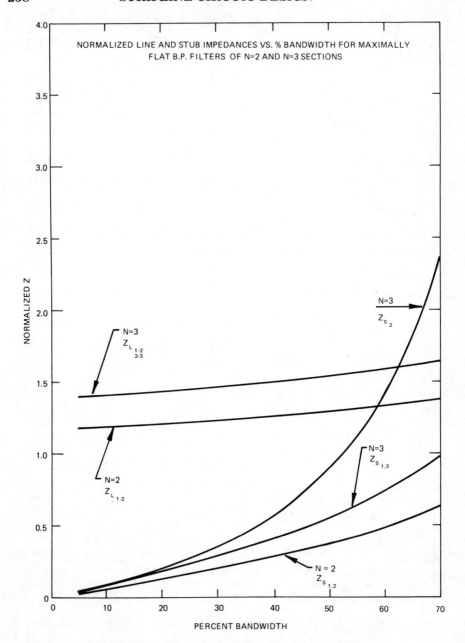

FIG. 7–21   Normalized Line and Stub Impedances vs. Percentage
Bandwidth for Maximally Flat Bandpass Filters Having
Two or Three Sections

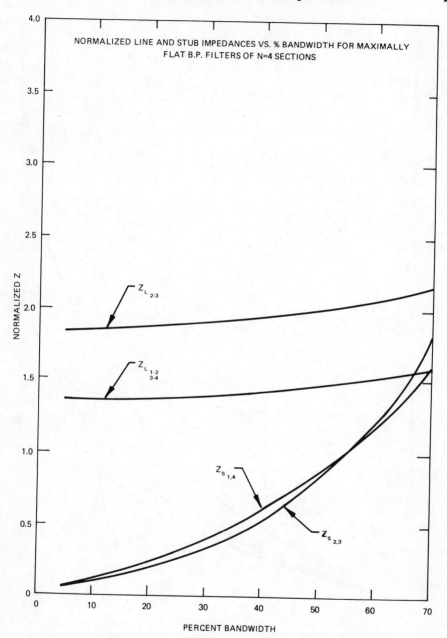

FIG. 7–22 Normalized Line and Stub Impedances vs. Percentage
Bandwidth for Maximally Flat Bandpass Filters Having
Four Sections

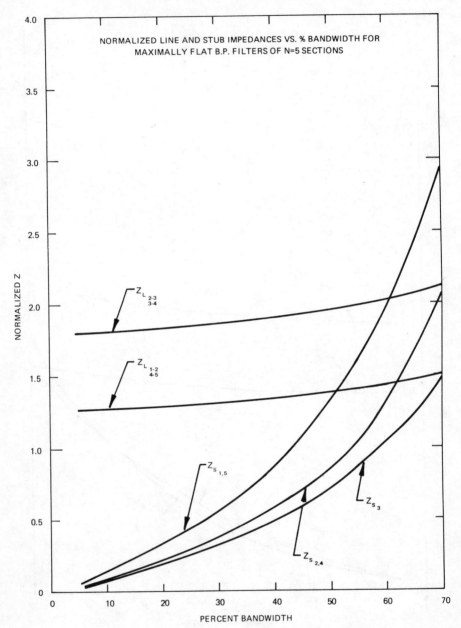

FIG. 7–23 Normalized Line and Stub Impedances vs. Percentage
Bandwidth for Maximally Flat Bandpass Filters Having
Five Sections

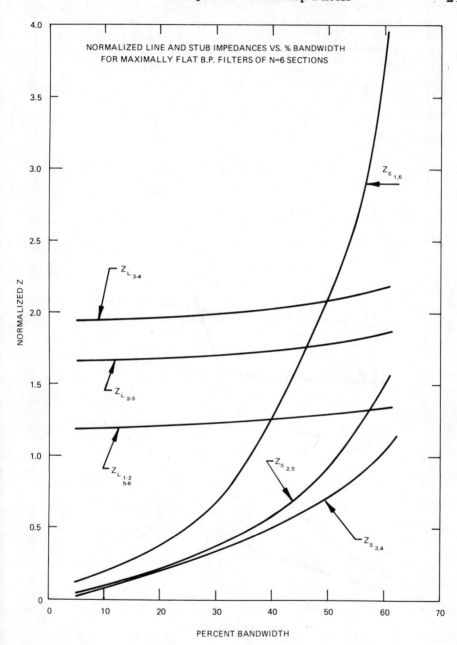

FIG. 7–24 Normalized Line and Stub Impedances vs. Percentage
Bandwidth for Maximally Flat Bandpass Filters Having
Six Sections

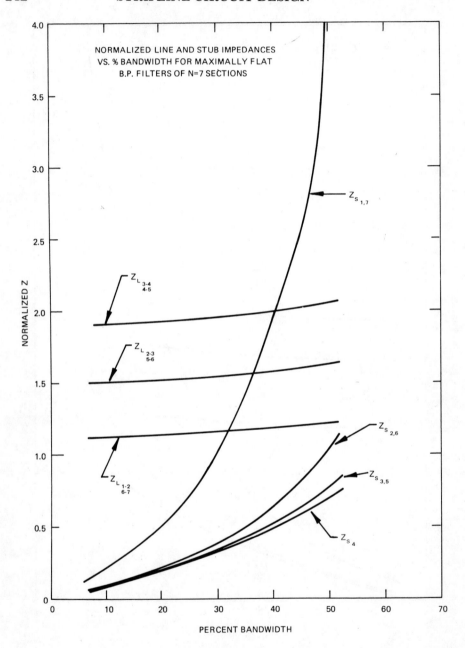

FIG. 7–25 Normalized Line and Stub Impedances vs. Percentage
Bandwidth for Maximally Flat Bandpass Filters Having
Seven Sections

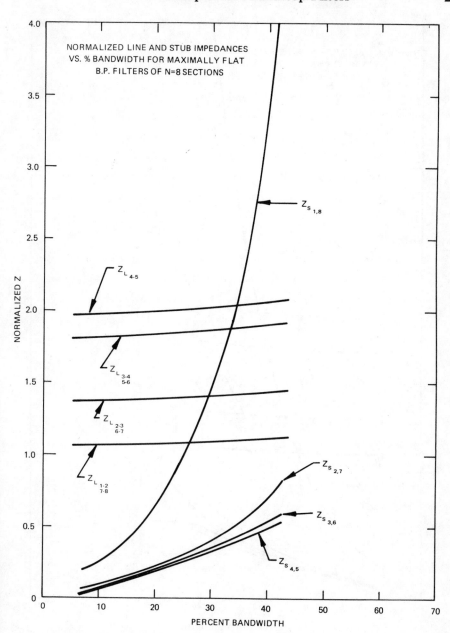

FIG. 7–26 Normalized Line and Stub Impedances vs. Percentage
Bandwidth for Maximally Flat Bandpass Filters Having
Eight Sections

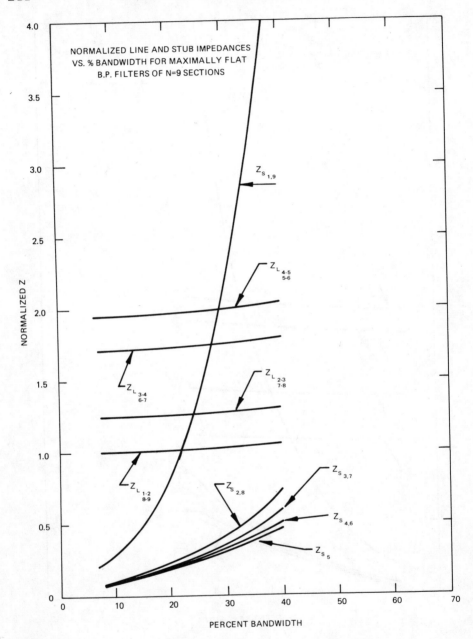

FIG. 7–27 Normalized Line and Stub Impedances vs. Percentage Bandwidth for Maximally Flat Bandpass Filters Having Nine Sections

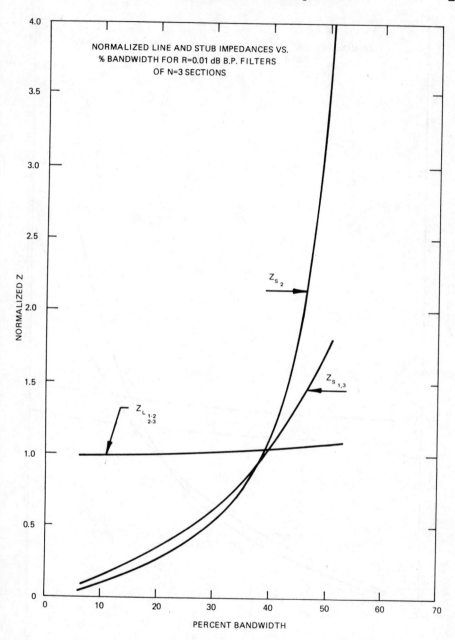

NORMALIZED LINE AND STUB IMPEDANCES VS.
% BANDWIDTH FOR R=0.01 dB B.P. FILTERS
OF N=3 SECTIONS

FIG. 7–28 Normalized Line and Stub Impedances vs. Percentage
Bandwidth for 0.01 dB Ripple Bandpass Filters Having
Three Sections

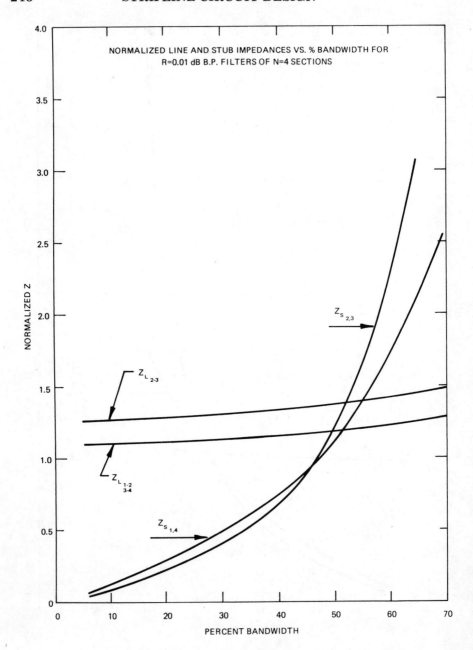

FIG. 7–29 Normalized Line and Stub Impedances vs. Percentage
Bandwidth for 0.01 dB Ripple Bandpass Filters having
Four Sections

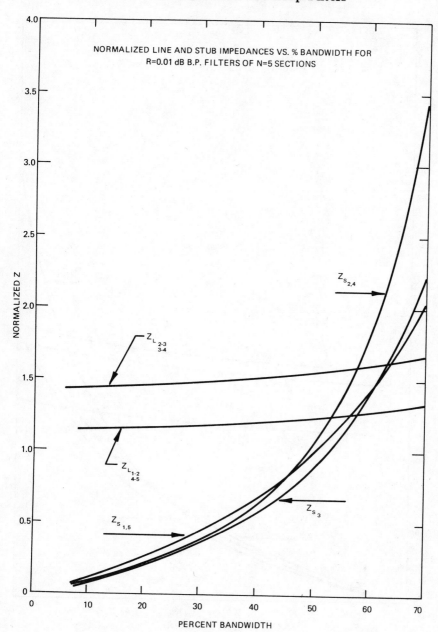

FIG. 7–30 Normalized Line and Stub Impedances vs. Percentage
Bandwidth for 0.01 dB Ripple Bandpass Filters Having
Five Sections

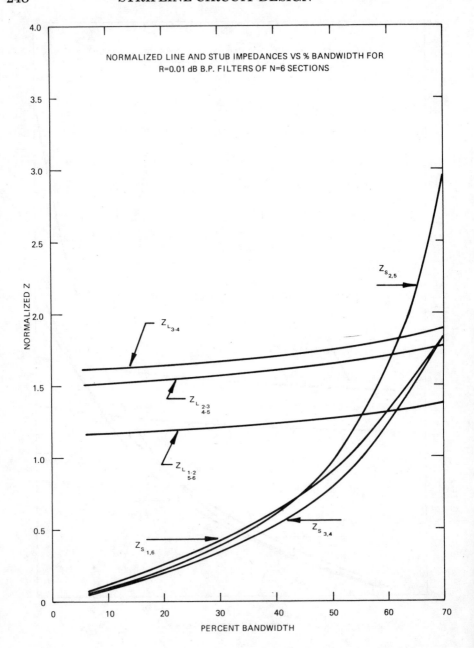

FIG. 7–31 Normalized Line and Stub Impedances vs. Percentages
Bandwidth for 0.01 dB Ripple Bandpass Filters having
Six Sections

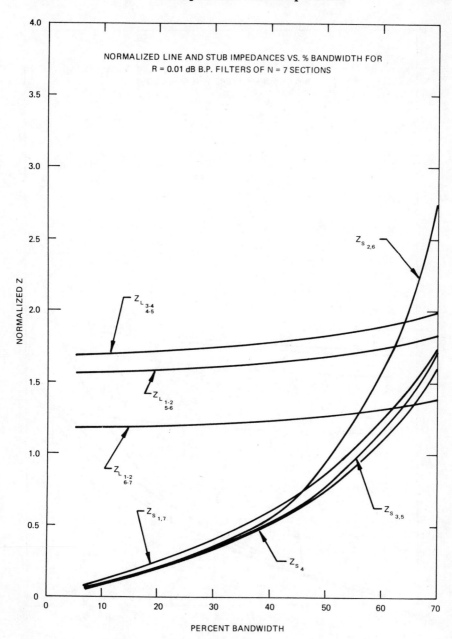

FIG. 7–32 Normalized Line and Stub Impedances vs. Percentage
Bandwidth for 0.01 dB Ripple Bandpass Filters Having
Seven Sections

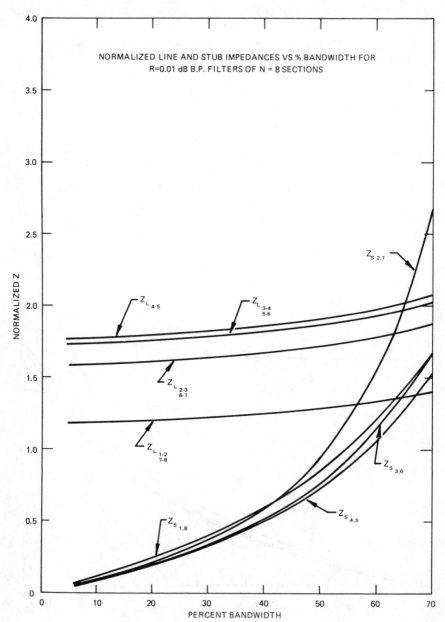

FIG. 7–33   Normalized Line and Stub Impedances vs. Percentage
Bandwidth for 0.01 dB Ripple Bandpass Filters Having
Eight Sections

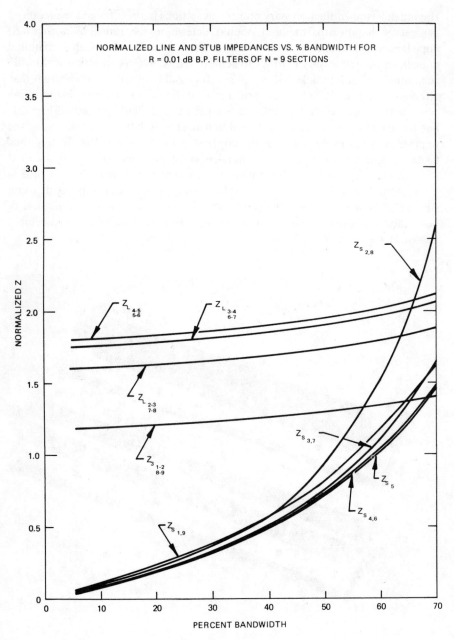

FIG. 7–34 Normalized Line and Stub Impedances vs. Percentage
Bandwidth for 0.01 dB Ripple Bandpass Filters Having
Nine Sections

Having determined the necessary number of sections in the filter and the required percentage bandwidth, the user should determine the range of normalized impedances and from this decide what transformations in either stub or mainline impedances are necessary in order to achieve reasonable construction within the constraints of solid dielectric stripline. It should also be remembered that although short-circuited lines are more difficult to achieve in stripline construction, open-circuited stubs have a tendency to "talk" to each other; this has the effect of reducing the achievable bandstop isolation and increasing the number of spurious responses in the upper stop bands of the filter. It may thus be desirable to attempt to use some short-circuited stubs and some open-circuited stubs in order to minimize these upper band spurious responses. Figure 7–35 is a photograph of a 2.5 GHz to 7.5 GHz thirteen-section stub filter built using single and double quarter-wave short-circuited sections. A filter making use of both short-circuited and open-circuited sections may be seen in the assembly shown in Figure 9–9.

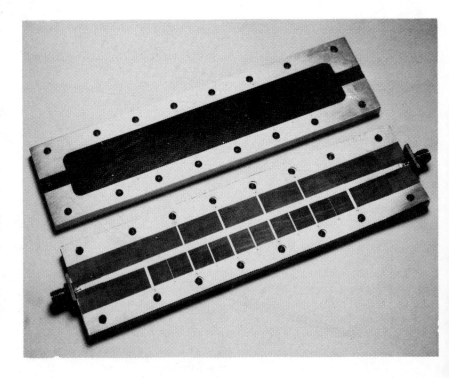

FIG. 7–35   Photograph of a 2.5 GHz to 7.5 GHz Maximally Flat
Shorted Stub Bandpass Filter

Traveling Wave Directional Filters

A unique type of bandpass/bandstop filter is shown in Figure 7–36. It consists of two main lines with coupled quarter-wave sections. The quarter-wave sections are interconnected by a ring whose mean circumference is one wavelength at the frequency of resonance of the filter[11] [12] [13]. A signal entering port A is coupled to port C at the frequency of resonance. All other frequencies are coupled to port B. Under normal conditions, port D is terminated; thus, the circuit from port A to B has a bandstop filter response and the circuit from port A to C has a bandpass filter response. In either case, if the unused ports are terminated, this provides a bandpass or bandstop filter which is matched at all frequencies in both the pass- and stop-bands. The device may be used as either a matched bandpass/bandstop filter or as a signal injection device for a broad-band frequency converter. In this case, a number of such filters connected to a common line with a semiconductor device may be used to inject several frequencies and to extract the sum or difference frequency after mixing. These filters may be designed in much the same way as a side-coupled directional coupler described by Chapters 4 and 5. The normalized even mode impedance for each section is given in equation 7–32, where the coupling characteristic is shown in equations 7–33, 7–34 and 7–35, with the parameters controlling them given in equations 7–36 through 7–38.

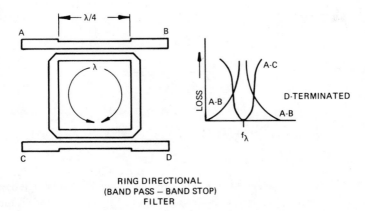

RING DIRECTIONAL
(BAND PASS — BAND STOP)
FILTER

FIG. 7–36   Traveling Wave or Ring Type Directional Bandpass-Bandstop
Filter

$$\text{norm } Z_{oe(k, k + 1)} \Big|_{k = o \text{ to } n} = \sqrt{\frac{1 + C_{k, k + 1}}{1 - C_{k, k + 1}}} \tag{7-32}$$

$$C_{01} = \sqrt{\frac{1}{\dfrac{(Q_e) A}{2 \pi m} + \dfrac{1}{2}}} \tag{7-33}$$

$$C_{(k, k + 1)} \Big|_{k = 1 \text{ to } (n - 1)} = m \pi Z_{(k, k + 1)} \tag{7-34}$$

$$C_{(n, n + 1)} = \sqrt{\frac{1}{\dfrac{(Q_e) B}{2 \pi m} + 1/2}} \tag{7-35}$$

$$(Q_e)_A = \frac{g_0 g_1}{\omega} \tag{7-36}$$

$$(Q_e)_B = \frac{g_n g_{n + 1}}{\omega} \tag{7-37}$$

$$Z_{(k, k + 1)} \Big|_{k = 1 \text{ to } n - 1} = \frac{\omega}{\sqrt{g_k g_{k + 1}}} \tag{7-38}$$

Curves of normalized even mode impedance versus percentage bandwidth are given for one-section and two-section rings in Figure 7–37 and Figure 7–38. The use of more than two sections is not recommended because of the extreme difficulty in building such structures. It is imperative in the construction of the ring-type directional filter that there be no backward wave to counter the effect of the forward wave in the ring. This means that the discontinuities of the ring must be kept at a minimum. This problem becomes more severe as the number

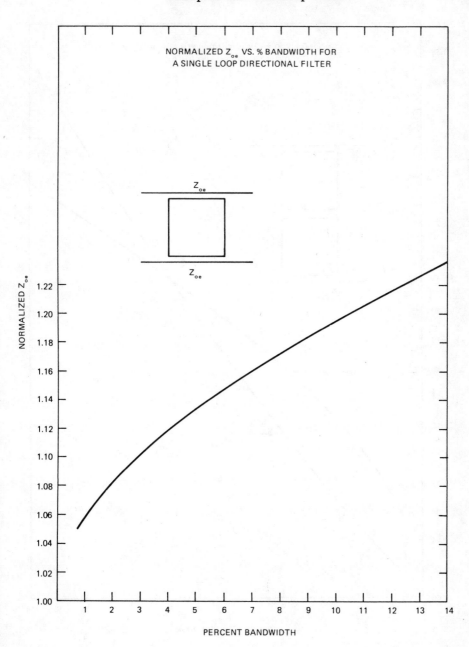

FIG. 7–37 Normalized Even-Mode Impedance vs. Percentage Bandwidth
for a Single-Loop Directional Filter

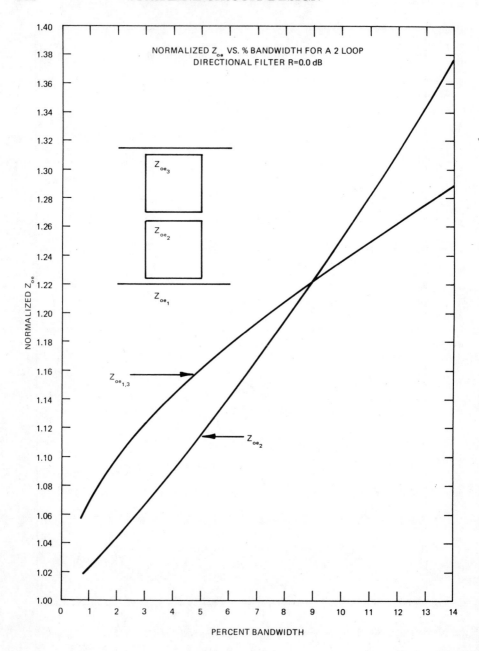

FIG. 7–38 Normalized Even-Mode Impedance vs. Percentage Bandwidth
for a Two-Loop Directional Filter having a
Maximally Flat Response

of rings is increased. In actual practice, it may be desirable to add tuning screws to each of the four sides of the ring to help minimize the reverse wave which may be generated due to coupler and corner discontinuities. The problem of internal ring VSWR's is serious[14] and should not be ignored. When properly constructed, however, the ring directional filter serves a function which cannot be equaled by any other simple device. As such, it remains a useful tool for the circuit designer.

Bandstop Filters

There are a number of types of bandstop filters, including the cavity resonator type, which is characteristic of waveguide construction; the capacitively coupled short-circuited stub type, which can be used in coaxial structures and stripline; and the coupled-section spur-line type shown in Figure 7–39.

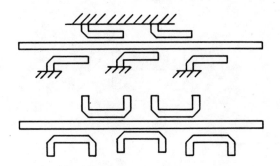

COUPLED RESONATOR BAND STOP FILTERS
WITH SHORTED OR OPEN RESONATORS

FIG. 7–39 Basic Configuration for Coupled Resonator Bandstop Filters having Shorted or Open-Circuited Resonators

This type of filter is the most useful for stripline construction because, once again, it makes use of the side-coupled configuration and can be realized by parallel-coupled line sections. Figure 7–39a shows the short-circuited version. Each section is one quarter-wave long. The short circuit shown after the right-angle at the end of each resonator should be as close as possible to the coupled section. Figure 7–39b shows the open-circuited version. In addition to the quarter-wave coupled section, an open-circuited eighth-wave section is added on each end. A quarter-wave section may be added to one end, but generally this requires more circuit area than the use of two eighth-wave sections. These open-circuit types are easier to construct, since the need to establish a known short-circuit is eliminated, but they have the same disadvantage as all open-circuit stub-filters: the achievable isolation is not as great, due to cross talk between the open-circuited stubs. The coupling dimensions of each section may be calculated from equation 7–44, where the value of voltage coupling is determined by equations 7–39 through 7–43.

$$\left. C_k \right|_{k \,=\, 1 \text{ to } n} = \sqrt{\frac{\dfrac{\pi}{2} \dfrac{1}{X(k)}}{1 + \left[\dfrac{\pi}{4} \dfrac{1}{X(k)} \right]}} \qquad (7\text{-}39)$$

where

$$X_{(1)} = \frac{1}{G_0 \, G_1 \, \omega} \qquad (7\text{-}40)$$

$$\left. X_{(k)} \right|_{k \,=\, 2,\, 4 \,\ldots\ldots\ldots} = Q \left[\frac{G_0}{G_k \quad \omega} \right] \qquad (7\text{-}41)$$

$$Q = \frac{1}{G_0 \, G_{n+1}} \qquad (7\text{-}42)$$

$$\left. X_{(k)} \right|_{k \,=\, 3,\, 5 \,\ldots\ldots\ldots} = \frac{1}{G_0 \, G_k \, \omega} \qquad (7\text{-}43)$$

$$\text{Normalized } Z_{oe_{(k)}} = \sqrt{\frac{1 + C_k}{1 - C_k}} \qquad (7\text{-}44)$$

Design curves have not been given for this class of filter since it is not commonly used. Most requirements for bandstop situations can be resolved by the use of bandpass, low-pass, or high-pass filters, thus making the bandstop filter, with its critical parameters and somewhat higher loss, less desirable as a structure, although it does have specific applications, particularly where narrow bands of frequencies need to be rejected in the middle of a broad spectrum of passband response.

REFERENCES

1   Cohn, Seymour, *Parallel Coupled Transmission Line Resonator Filters,* MTT-6, No. 2, April 1958, pp. 223–231.

2   Smith, H., *Computer Generated Tables for Filter Design,* Electronic Design, Vol. 11, May 1963, pp. 54–57.

3   Richardson, J. K., *Gap Spacing for End-Coupled and Side Coupled Stripline Filters,* MTT-15, No. 6, June 1967, pp. 380–382.

4   Matthei, Jones, & Young, *Microwave Filters, Impedance Matching Networks and Coupling Structures,* McGraw Hill, New York, 1964, pp. 472–476.

5   Richardson, J. K., *Gap Spacing for Narrow Bandwidth End Coupled Symmetric Stripline Filters,* MTT-16, No. 8, August 1968, pp. 559–560.

6   Cohn, op. cit.

7   Matthei, Jones, & Young, op. cit., pp. 440–442.

8   Oliner, A. A., *Equivalent Circuits for Discontinuities and Balances Strip Transmission Line,* MTT Vol. 3, March 1955, pp. 134–143.

9   Altschuler, H. M., and Oliner, A. A., *Discontinuities in the Center Conductor of Symmetric Strip Transmission Line,* MTT, Vol. 8, May 1960, pp. 328–339.

10   Cristal, E. G., *New Design equations for a Class of Microwave Filters,* MTT, Vol. 19, No. 5., May 1971, pp. 486–490.

11   Coale, F., *A Traveling Wave Directional Filter,* MTT-4, October 1956, pp. 256–260

12   Cohn, S. B., Coale, F., *Directional Channel Separation Filters,* Proceedings IRE, No. 44, 1956, pp. 101.

13   Matthai, Jones & Young, op. cit., pp. 867–872.

14   Adams, D. K., and Weir, W. B., *Wideband Multiplexers Using Directional Filters,* Microwaves, Vol. 8, No. 5., May 1969, pp. 44–50.

15   Cristal, E. G., *Bandpass Spur Line Resonators,* MTT-14, No. 6, June 1966, pp.296–297.
Cristal, E. G., Correction to, *Bandpass Spur Line Resonators,* MTT-14, No. 9, September 1966, pp. 436.

CHAPTER
—8

Mixers, Switches, and The Application of Hybrids

The application of hybrids to mixers, switches, discriminators, and other circuits is of primary importance in component and subsystem design. Of the numerous hybrids available, three major varieties are most commonly used in stripline design.

The first of these, the 90° hybrid, is, perhaps, the most useful of all types. This is shown in Figure 8–1a.

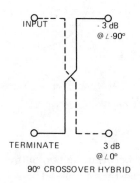

FIG. 8–1 Basic Hybrid Configuration

In its normal configuration, it consists of a cross-over constructed with co-linear output arms, so that it may be used in some of the circuit configurations which will be described. An input signal is split equally to the two output arms with a phase quadrature relationship between the outputs. Both the quality of the hybrid, and the mismatches on the two output arms determine the amount of power which is shunted into the terminated arm. For the perfect case, there is little or no power in the terminated arm; however, under practical conditions, power at this port is typically somewhere between 10 and 20 dB down from the input power.

  The second hybrid, the in-phase, in-line three-port type of hybrid, is shown in Figure 8–1b.

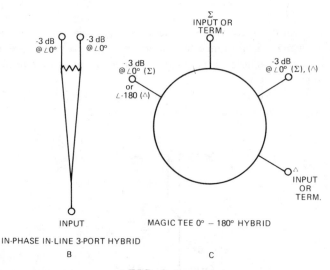

FIG. 8–1

An input signal splits equally between the two outputs with an equal phase and amplitude relationship between the outputs. The reverse isolation between the two outputs is typically in the order of 20 dB and may vary between 14 dB and 35 dB as a function of frequency and the design of the actual hybrid.

  The third type of hybrid, a four-port magic-tee type known as the rat-race, is shown in Figure 8–1c. The co-linear arms have equal amplitude outputs from either the sigma or delta input arms. In the case of an input from the sigma arm, both co-linear arm outputs are in phase, while for an input from the delta arm, the two co-linear output arms are out of phase by 180°.

Signal Addition or Cancellation

Two signals of equal amplitude may be added or combined by use of 0°, 90°, or 180° types of hybrids as a function of their required phase relationship. For most cases, 0° or 180° hybrids are the preferred types. Figure 8–2 is a curve of insertion loss versus phase unbalance for equal amplitude signals entering two ports of a 0° or 180° type hybrid. There is a reasonably large phase tolerance around the 0° optimum condition, so that errors as great as 20° or 30° will result in only a small fraction of a dB loss of the combined signal. A similar tolerace exists for amplitude, for a very little loss resulting from an amplitude unbalance as great as 0.2 or 0.3 dB.

The cases of signal cancellation and image rejection are shown in Figure 8–3.

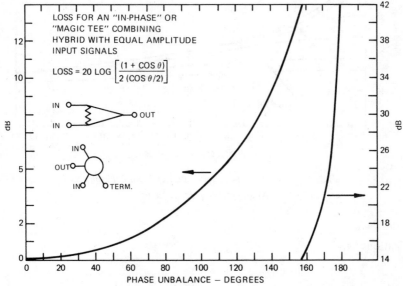

FIG. 8–2   Loss vs. Phase Unbalance for a Combining Hybrid
with Equal Amplitude Input Signal

Here, the depth of null, or rejection, for signals which are nominally 180° out of phase is plotted for a variety of amplitude unbalances versus phase error in degrees from the ideal 180° cancellation case. As would be expected for extreme depths of null, almost perfect amplitude and phase balance are required.

However, for levels in the 20 dB to 30 dB region, substantial latitude is also possible in both the amplitude and phase accuracy of the signals to be nulled. These curves may be applied to the depths of null in such circuits as monopulse comparators and interferometers, as well as to the isolation which can be achieved in image rejection, or image separation, mixer circuits.

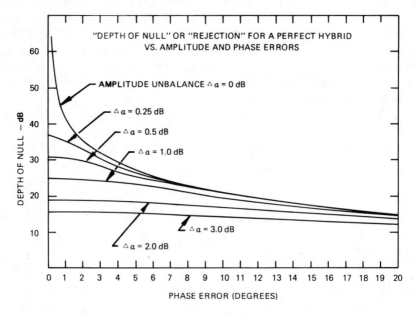

FIG. 8-3   Depth of Null for a Perfect Hybrid Plotted vs.
Amplitude and Phase Errors

Applications

The 90° hybrid is most frequently applied in two circuit configurations. One of these is for balanced mixer use; the other is for reduction of VSWR when reflective devices must be employed in a circuit. Figure 8-4 shows how equal reflective devices can be placed between two 90° 3.0 dB hybrids in order to maintain both the input and output match under all reflection conditions.

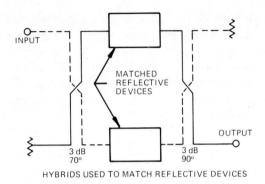

HYBRIDS USED TO MATCH REFLECTIVE DEVICES

FIG. 8–4   Hybrid Layout Configuration used to Match
Reflective Devices

The signal enters the first hybrid which splits it with an equal amplitude and 90°
phase relationship. It passes through the mismatched devices and recombines in
the second hybrid, to emerge at the output port. Reflections which may occur
from the two inserted devices will combine and be shunted to the termination
on either the input or output hybrid, depending upon the source of the signal.
These conditions will hold true for up to 100% reflection from both devices. This
basic circuit configuration is one of the most widely used for amplifier matching,
PIN diode switch matching, double-balanced mixer matching, and diplexing. It
is also used in similar applications where reflective devices must be utilized in
the circuit, and where inherent mismatches cannot be tolerated because of their
interaction with other circuit elements.

The circuit may also be used without any devices whatever between the
two 3.0 dB hybrids. In this case, a 0 dB coupler configuration exists as shown
in Figure 8–5.

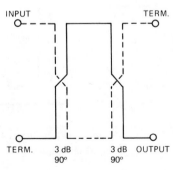

FIG. 8–5   0 dB Coupler Configuration

This seemingly useless circuit has many useful applications. Among these is the "cross-over". A signal entering the circuit from the upper left-hand side will emerge at the lower right-hand side, and a signal entering from the lower left-hand side will emerge at the upper right-hand side. Thus, the 0 dB coupler configuration can be used to permit two lines to cross each other with typically 15 dB to 20 dB isolation between those two lines. It is also useful in extremely broadband balanced mixers for IF extraction or dc bias injection.[1]

Instantaneous Frequency Monitors

An instantaneous frequency discriminator, or microwave interferometer, may be made from several microwave hybrids and a delay line, using the circuit in Figure 8–6.

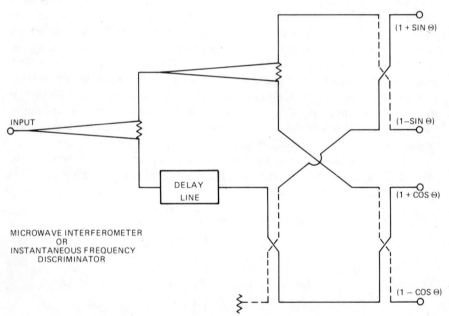

FIG. 8–6   Microwave Interferometer or Instantaneous Discriminator Circuit

The input signal is split between the two channels, one of which contains a delay line feeding a matrix of one 0° and three 90° hybrids. The outputs of this circuit are four signals whose amplitude vary as (1 + Sin θ), (1 − Sin θ), (1 + Cos θ), and (1 − Cos θ) where the angle θ is a function of frequency and the length of the delay line. If these signals are detected and compared, they may then be used to determine the frequency of an unknown signal entering the circuit.[2] The display mechanism may be an oscilloscope where the output from one differential amplifier feeds the vertical coordinates and the output from the other differential amplifier feeds the horizontal coordinates, thus creating a display where the angular position of rotation is a measure of the unknown frequency.[3] These outputs may also be fed through an analog-digital converter into a digital computer for further system processing. This type of circuit has become extremely useful and widespread in ELINT and ECM applications. It is normally built to cover single-octave bandwidths rather than multi-octave bandwidths, since the second harmonic signals can create ambiguous responses.

Power Dividers, Multicouplers, And Antenna Matrices

Cascades of 0° or 90° hybrids are frequently used for power division networks. In their simplest form, binary numbers or outputs are used so that even split hybrids can be applied. There are, however, requirements where a non-binary split may be necessary. In these cases, equal-split hybrids can be combined either with unequal-split hybrids or with conventional directional couplers, depending upon the phase requirements of the power division network. Even low frequency power dividers can be built using this technique. Multi-couplers operating as low as the UHF and VHF bands can be constructed by meandering the lines. The many amplitude and phase slope requirements of antenna feed systems can also be satisfied by the proper combination of hybrids and couplers. The use of iris-coupled elements for antenna arrays is greatly simplified when the feed structure is built using these techniques.

Diplexers

Diplexers may be constructed using matched pairs of conventional bandpass filters mounted between 90° hybrids as shown in Figure 8–7.

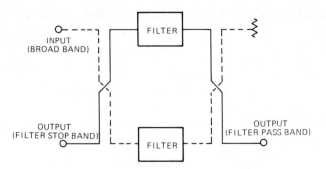

FIG. 8–7   Hybrid Circuit Used for Diplexing or Multiplexing

A signal entering at the upper left hand corner of the diagram will pass through the filters and recombine at the lower right hand corner for frequencies within the passband of the filter. All reflected signals from the filter will exit at the output port at the lower left-hand corner of the circuit. This port may be used for further processing circuits, or may be fed to another diplexer built using the same configuration as shown in Figure 8–7, thus permitting channel dropping to be achieved by the cascade of a number of these circuits.

As in the case of other matched reflective devices between hybrids, the input port is matched at all frequencies. This device is useful for circuits which require wider or greater skirt selectivity than that achievable by the use of ring directional filters. It should be remembered, however, that the rejection at the reflected output port is no greater than the inherent isolation of the 90° hybrid modified by the VSWR of the matched filters' passband response.

Diode Mounting Configurations and IF Filtering Circuits

Basically, two types of hybrids are applied to the balanced mixer, the 90° quadrature and the 180° magic tee. If all conditions were perfect, there would be little difference between these two types of hybrids. In practice, it is virtually impossible to match the mixer diodes perfectly at all frequencies and under all LO drive conditions. Therefore, some degradation in the hybrid performance can be expected as a result of the reflections from these diode mounts. In the case of the 90° hybrid mixer, the input VSWR at either port will generally be low since the reflections from the mixer diodes will be shunted out the opposite port. The isolation between the signal port and LO port will be strictly a function of the return loss of the diode mount. Despite the fact that the isolation is low, the AM noise cancellation, which is a function of the amplitude and phase balance of the hybrid itself, and not of the return loss of the diodes, is generally as high with a 90° type as it is with the magic-tee type. These factors, combined with the ability to operate over octave and multi-octave bandwidths with ease, make the 90° hybrid the most frequent choice for broadband mixer applications. Typical performance characteristics for a 90° mixer are: signal to LO isolation varying between 6 and 20 dB as a function of match and bandwidth; VSWR's in the vicinity of 1.5; and noise figures which are largely a function of the diode choice.

The magic-tee, on the other hand, is a circuit which provides inherently high isolation between the signal and LO ports regardless of the mismatches of the diode mounts, provided those mismatches are balanced. Typical performance for a magic tee mixer would be 20 dB minimum LO to signal isolation; VSWR in the order of 1.5 to 2.5 as a function of the diode mismatch, and once again, noise figures in the same order of magnitude as that available for the diodes themselves. Naturally, as bandwidths increase and the circuit complexity of the device increases, noise figure will be degraded accordingly.

Mixers

The basic circuit of a mixer, or frequency converter, consists of a microwave structure which combines two signals feeding them into either one, two, or four diodes whose non-linear characteristics provide mixing or frequency conversion. A typical configuration for a balanced mixer is shown in Figure 8–8.

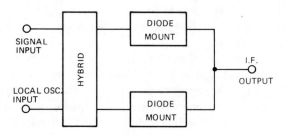

FIG. 8–8   Simple Balanced Mixer Circuit

One degree of freedom which exists for the designer is the choice of series or shunt diode mounts. A series diode configuration, which is shown in Figure 8–9a, is one in which the diode has a dc and IF return at the RF input side of the diode and an RF bypass capacitor at the IF output side of the diode.

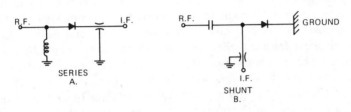

FIG. 8–9 Series and Shunt Diode Mount Configuration

This output capacitor may be either a lumped type or a filter structure designed to stop the RF and LO frequencies, thus providing a return to ground for these signals and establishing the necessary current loop. Similarly, the IF frequency generated as a result of the mixing action has its return to ground on the RF side and passes through the filter into the next stage of the assembly, which is usually a preamplifier.

The shunt diode configuration is shown in Figure 8–9b and consists of a diode which is mounted from the main RF line and IF line to ground.

A dc and IF block must be provided on the RF side of the structure. It is also necessary to provide RF filtering for the IF side of the diode mount. The choice of a series or shunt configuration for the diode mount is a function of the RF frequency and bandwidth, the IF bandwidth, and any possible mechanical requirements, such as the need for replaceable diodes. It is possible to have replaceable diodes in a series configuration; however, they generally lend themselves more readily to the shunt configuration, since one end of the diode is grounded. This grounding can be conveniently done with a removable cap, thus permitting replacement of the diode. A replaceable diode mount is shown in the photograph of Figure 8–10.

Another consideration of importance is the low frequency response of either a dc return or an IF and dc block. These structures must be chosen to be compatible with the low frequency responses of the chosen IF frequencies and bandwidths. Dc returns are generally made using quarter-wave short-circuited stubs on the RF side of the diode. It is important that the low-frequency isolation of these structures be high enough so that the resultant insertion loss at the IF

FIG. 8–10   Photograph of Removable Diode Mounts used for Miniature
Cartridge Mixer Diodes

frequencies is low. Figure 8–11 shows the main-line isolation on a 50 ohm line for a short-circuited quarter-wave stub of varying impedance, normalized to the design frequency for which the stub was constructed. Similarly, Figure 8–12 shows the low frequency response for a single-section series stub which can be used in the shunt diode configuration to prevent RF and dc leakage into the RF circuitry. When the shunt configuration is used, it is necessary to extract the IF frequency through an RF bandstop filter. Two simple filters are shown in Figure 8–13 and 8–14 as a function of varying shunt stub impedance. For most mixers with RF bandwidths of up to an octave, these structures are satisfactory. In the case of multi-octave mixers, however, it may be necessary to go to more sophisticated filter structures, making use of the high-pass and low-pass multi-section circuits described in Chapter 6.

Whether the series- or the shunt-diode configuration is used, several types of mixer circuits are possible. A single-ended mixer as shown in Figure 8–15 is

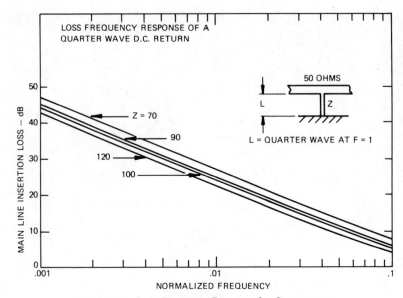

FIG. 8–11   Low Frequency Response for Quarter-wave
dc Return

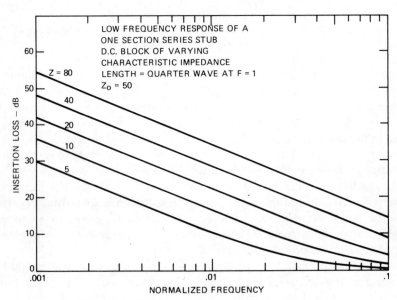

FIG. 8–12   Low Frequency Response for One-section Series-stub
dc Block for Several Characteristic Impedances

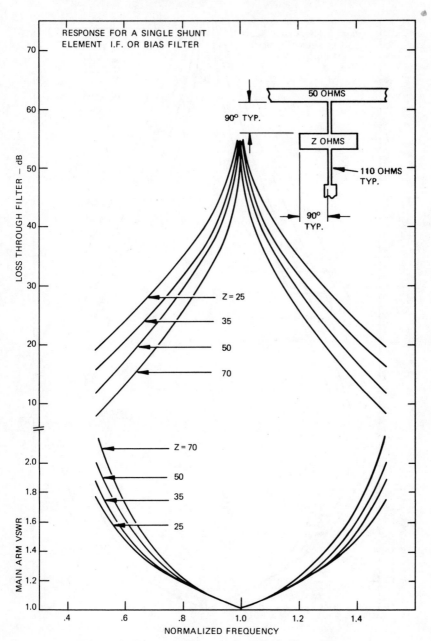

FIG. 8–13 Frequency Response for a Single Shunt Element
IF or Bias Filter

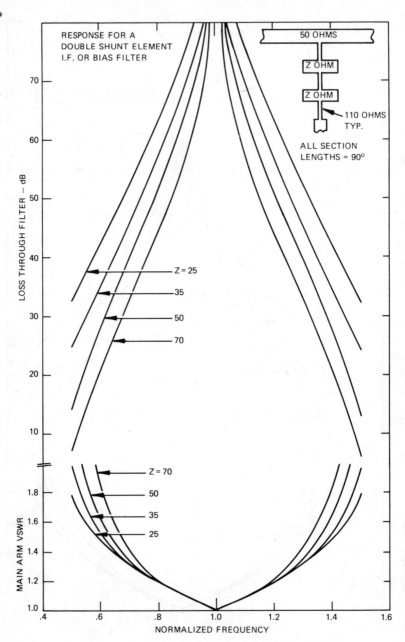

FIG. 8–14 Frequency Response for a Double Shunt Element
IF or Bias Filter

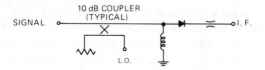

FIG. 8–15   Single-Ended Mixer Configuration

most conveniently made using a directional coupler. The signal is applied on the main arm of the coupler with the LO being coupled by means of the coupling section. No AM noise cancellation will be achieved, unlike the case of the balanced mixer, although FM noise cancellation will occur. More LO power is necessary than usual as a result of the coupling factor of the directional coupler, which should be kept as loose as possible in order to minimize main line losses. This type of mixer is most frequently used in AFC applications, where signal levels and LO levels are high and noise figure is not a serious consideration. The balanced mixer in Figure 8–8 is the most universally applied type of mixer. It is useful because of its AM noise cancellation, its generally low noise figure, its LO to signal isolation, and its simplicity of construction. An additional type of balanced mixer is the well-known doubly-balanced mixer which is most frequently constructed using toroidal transformers. Its stripline counterpart is the double-balanced mixer shown in Figure 8–16.

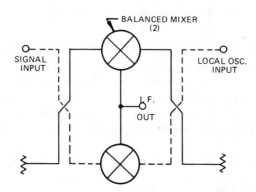

FIG. 8–16   Pseudo Double-Balanced Mixer

While several types of true doubly-balanced microwave mixers exist, including the Balun-ring[4] and star configurations[5 6], the type most frequently used for planar stripline construction is the pseudo double-balanced mixer. This consists of two conventional balanced mixers built using either 90° or 180° hybrids placed between two 90° hybrids. Thus, the signal input splits to both balanced mixers as does the local oscillator input. Because of the circuit configuration, the LO and signal leakages through the balanced mixers are shunted to the isolating hybrids' terminated ports. Similarly, the reflections from the two balanced mixers are shunted into the hybrid loads, thus providing a mixer which has good LO to signal isolation and good VSWR at all ports.

Another two-mixer configuration between hybrids is the image separation or image rejection mixer shown in Figure 8–17.

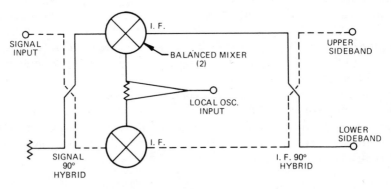

FIG. 8–17   Image Rejection Mixer Assembly

Here, two balanced mixers are fed by a 90° hybrid on the signal input with a 0° hybrid on the LO input. The IF outputs are fed through an IF frequency 90° hybrid which separates the upper side-band frequencies from the lower side-band frequencies. The degree of separation, or image rejection, can be determined as a function of the phase and amplitude errors from the curve in Figure 8–3. This type of mixer is particularly useful when significant noise power exists in the image band.

Multiplexer Type Mixers

Multiplexers can be used for mixers or frequency converters and are particularly useful where the frequencies involved are widespread and the intermediate frequencies are high. This type of mixer is shown in Figure 8–18.

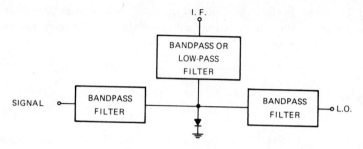

FIG. 8–18   A Multiplexer Mixer Configuration

An input signal filter and an input LO filter allow the two signals to be combined in the mixing diode. The intermediate frequency is extracted through the IF frequency filter. Naturally, it is necessary that none of the frequency bands overlap. It should also be recognized that the conversion loss of such a circuit is degraded by the insertion loss of the filters at the inputs, as well as at the outputs, and that these losses are generally greater than simple hybrid losses. For narrow band applications, image enhancement techniques can sometimes overcome a portion of these losses.

Choices of Diodes for Mixers

There are three basic diode types which are applicable to a mixer design, whether it be balanced, single ended, or image rejection. These are the point-contact diode, the back diode and the Schottky diode. The point-contact is the oldest and most established diode. Most of these diodes are made of silicon, although some early models were made of germanium. Their RF impedance is generally higher than that of other diode types, which makes their matching somewhat difficult over broad bandwidths. Their major virtue comes from the fact that they are readily available and generally inexpensive, except for the very lowest noise figure types. For cases where low IF frequencies are being used, the 1/F or low frequency "flicker noise" contribution of these diodes is high;[7] thus, for audio frequencies and "0" IF, the point-contact is not the ideal choice. Additional problems also accrue from the fact that the point-contact is exactly what its name suggests: a point contact which is subject to changes in characteristics as a result of vibration, shock, and acceleration, thus reducing its reliability. However, in the case of pulsed radars without limiter protection, the widely used point-contact diode has actually been shown to be more resistant to burnout than other types.

Back Diodes

The back diode, which is a variation of the tunnel diode, is generally made of germanium. Back diodes have a low RF impedance which makes them readily matched over a broad frequency range. Their 1/F low-frequency noise characteristics are the best of all diodes, but they are not generally used in mixer applications. Their major drawback is their low level of saturation, which reduces their dynamic range.

Schottky Diodes

The diode most used for new mixer development is the Schottky diode. Schottky diodes have come into major usage for mixer applications in the last eight years. It is almost universally made of silicon, although some gallium arsenide Schottky's are now available. The impedance of Schottky diodes is more dependent upon LO drive level than is the impedance of either the back diode or the point-contact diode. Thus, with proper LO drive, it is possible to match a Schottky diode over an extremely broad bandwidth, up to and including 20:1. The 1/F low frequency characteristics are on a par with the back diode which makes it extremely useful for homodyne systems. Since the Schottky diode is a junction diode, its reliability is increased by improved mechanical construction, although it is more susceptable to pulse burnout than are other mixer diodes.

An added advantage to the use of the Schottky diode comes from the application of dc bias. Dc bias not only improves the match of a Schottky device over a broad bandwidth, particularly in the higher frequency ranges, but also permits its operation in a "starved" LO. Thus, while it may normally be necessary to have 0 dBm to +3 dBm of LO drive power per diode in a balanced mixer, dc bias will permit starved LO operation at levels as low as −7 to −10 dBm per diode with only a small degradation in noise figure and an actual improvement in match conditions. Naturally, intermodulation products and saturation levels will be the same as if the LO drive were unsupplemented by dc power.

Noise Figure

A conventional singly-balanced mixer is capable of a single sideband noise figure varying between 6.0 dB at the low end of the frequency range and up to 7.5 or 8.0 dB in the Ku-band frequency range. This noise figure is measured with a 1.5 dB IF at some nominal IF frequency such as 30 or 60 MHz. Image rejection mixers have noise figures approximately 1.0 dB higher than the conventional mixer due to the circuit losses of the input and IF hybrids. The IF hybrid losses can be minimized by the use of dual preamplifiers, in which case an image rejection mixer assembly can be built whose noise figure is within 0.5 dB of the individual balanced mixer used to construct it.

Image enhancement has been applied over narrow bands to balanced mixers by a number of people.[8][9] It involves the technique of reactively terminating the image frequencies and harmonic frequencies which are generated in the diode. At this time, image enhancement mixers in S-band with noise figures of approximately 5.5 dB have been built on a production basis. Others have reported laboratory noise figures even lower.

Doubly-balanced mixers in low frequency ranges are also available with noise figures in this order of magnitude. A novel combination of doubly-balanced techniques and image enhancement mixers has been built in suspended substrate microstrip.[10] While not directly applicable to stripline construction, its performance characteristics, which include noise figures in the 4.0 dB region, are dramatic enough to warrant its consideration for possible integration into stripline packages. Its construction is unique, however, and not characteristic of the typical double-balanced mixer built in stripline. The same additional insertion losses as the image rejection mixer and therefore, in general, would not have as good a noise figure as a simple singly balanced mixer if built in stripline.

IF Bandwidth

The IF bandwidth is a function of both the RF frequency of the mixer and its filter circuits, as well as whether or not the mixer has been constructed as a series, shunt, or true doubly-balanced configuration. The series diode configuration has the advantage of making use of a dc return on the RF side of the mixer which generally will have a wide band response at the IF frequencies, and a low-pass filter response on the IF end of the mixer diode, which can be built to pass relatively high IF frequencies. The shunt diode configuration on the other hand has the advantage of providing a short circuit at one end of the diode whose location is identical for all frequencies, RF, LO and IF. It requires, however, that good quality high-pass filters and low-pass filters be built into the RF circuit for the extraction of the IF frequency.

Stripline mixers are commonly built with IF frequencies and bandwidths going up to several hundred MHz, with 500 MHz being a reasonable upper limit for conventional design. Special mixers with IF frequencies as high as 0.5 times the lower edge frequency of the RF bandwidth are also available. In general, however, IF frequencies above several hundred MHz require special techniques, thus increasing the complexity and cost of the mixer.

Isolation

Balanced mixers have LO to RF isolations which are a function of their circuit type. For the 90° balanced mixer, isolation may range anywhere from 20 dB to 6.0 dB for reasonable mixers and is a function largely of the diode mismatch, since the isolation is directly related to the return loss and balance of the diodes. The magic-tee mixer, however, will provide higher isolation, because the diode mismatch is manifested as a high signal and LO port VSWR. Doubly-balanced mixers have an inherently higher degree of LO to RF isolation as a result of their dual-hybrid construction. Typically, minimum levels of isolation equal to 20 dB can be anticipated, with even higher isolation over narrow frequency ranges.

VSWR

The VSWR of balanced mixers is a function of the choice of 90° or 180° hybrids for their construction. Typically 90° hybrids will provide VSWR levels over octave and multi-octave bands in the vicinity of 1.5 with multi-octave band construction having peak values in the order of 2:1. Magic-tee mixers, however, have higher VSWR's which may range from 2:1 to 3:1. As in the case of isolation, the doubly-balanced mixer solves the problem of reduced VSWR as well as low isolation. Doubly-balanced mixers can be typically expected to operate at VSWR's less than 1.5 and at isolations greater than 20 dB.

Intermodulation Products

The subject of intermodulation products and balanced mixers has increased in significance in recent years. These intermodulation product levels are a function of mixer construction, but more notably of LO drive levels and semiconductor properties. Thus, no consistent guidelines can really be established which will accurately predict the general case of intermodulation products. Figure 8–19 is a plot of the intermodulation response of an octave-bandwidth 90° hybrid, C-band balanced mixer and will serve to provide rough guidelines in the area of intermodulation products.

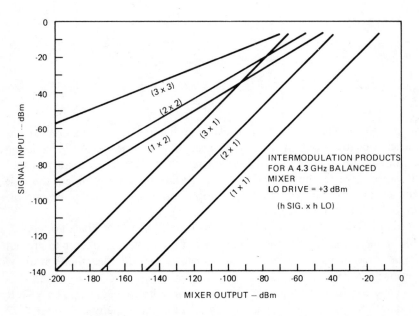

FIG. 8–19 Intermodulation Curve for a 90° Hybrid
C-band Balanced Mixer

Diode Switches

In much the same way that the Schottky and point-contact diodes can be incorporated into stripline circuits to provide mixers and frequency-converter functions so the PIN diode can be used to perform microwave switching. This diode has the characteristics of an extremely low-loss resistor when forward-biased and an extremely high resistance in parallel with a very small capacitance in the zero- or reversed-bias condition. The equivalent circuit for a PIN diode in both bias conditions is shown in Figure 8–20.

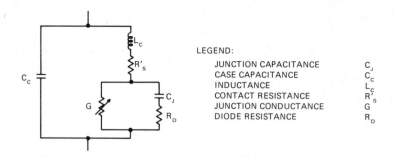

LEGEND:

JUNCTION CAPACITANCE	C_J
CASE CAPACITANCE	C_C
INDUCTANCE	L_C
CONTACT RESISTANCE	R'_S
JUNCTION CONDUCTANCE	G
DIODE RESISTANCE	R_D

FIG. 8–20   PIN Diode Equivalent Circuit

If the parasitic values of the diode package are ignored, as can be done at low frequencies, or tuned out, as can be done at high frequencies, then the diode can be considered to be a very small resistance in the forward bias case, or a very small capacitance in the reverse bias case. While inaccurate, unless the case parameters are taken into account, this simplification can be used to explain the operation of a PIN diode switch.

Figure 8–21a shows the shunt configuration for a PIN diode switch.
In this application, the diode is mounted from the center conductor to ground where appropriate bias lines and dc blocks permit bias in either the forward or the reverse direction.

In the reverse direction, the diode shunt capacity will show up as a VSWR on the line. This can be seen for several values of shunt capacity in Figures 8–22 and 8–23.

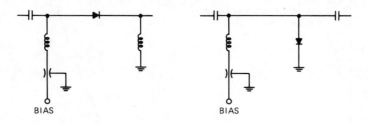

FIG. 8–21   Series and Shunt Mounted Configuration for
PIN Diode Switches

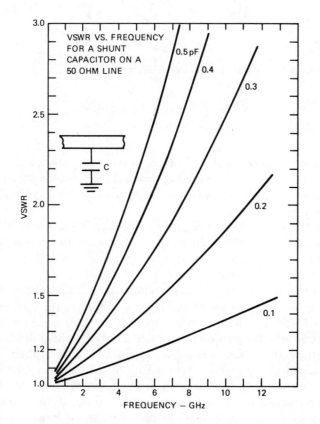

FIG. 8–22   VSWR vs. Frequency for a Shunt Capacitor on
a 50 Ohm Line

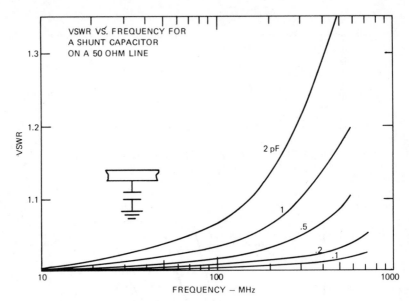

FIG. 8–23   VSWR vs. Frequency for a Shunt Capacitor on
a 50 Ohm Line

When the diode bias is changed to forward, or conducted, bias, the low shunt resistance of the diode causes a large reflection on the line resulting in a relatively high degree of isolation. This can be seen in Figure 8–24 for a single resistance to ground for a 50 ohm line.

It should be noted that although it is theoretically possible to achieve 20 to 30 dB of isolation with a single shunt resistance to ground, parasitic inductances relating to the diode structure generally limit this to 20 dB or slightly less for a single diode mounted in shunt on a 50 ohm line.

As an alternative method of construction, the diode can be mounted in series on the 50 ohm line as shown in Figure 8–21b. Here the bias conditions will be reversed from the shunt situation; forward bias will cause low insertion loss while reverse bias will provide attenuation. Unlike the shunt diode case, the reverse bias attenuation is far more frequency sensitive, since as the isolation is directly a function of the reverse bias capacity of the diode as related to the frequency of operation. This can be seen in Figures 8–25 and 8–26. Again, it should be remembered that unless the case parasitics are tuned out, or are so small they can be ignored, these figures will be degraded. When forward-biased, the loss of the line will be a function of the R_s of the diode and this, as well as its VSWR, can be seen from Figure 8–27.

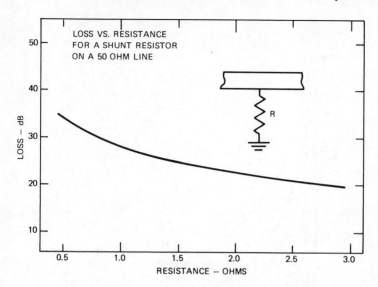

FIG. 8–24   Loss vs. Resistance for a Shunt Resistor on a
50 Ohm Line

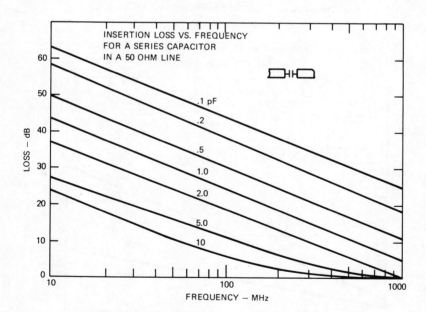

FIG. 8–25   Insertion Loss vs. Frequency for a Series Capacitor
on a 50 Ohm Line

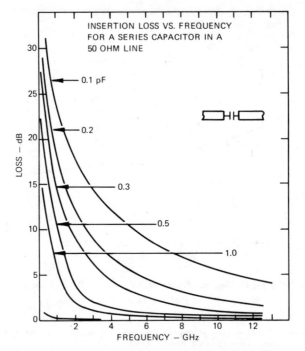

FIG. 8–26   Insertion Loss vs. Frequency for a Series Capacitor
in a 50 Ohm Line

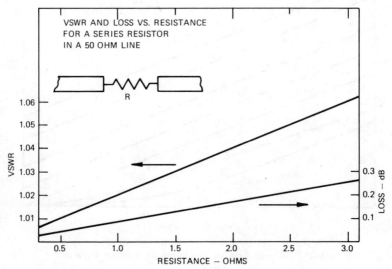

FIG. 8–27   VSWR and Loss vs. Resistance for a Series Resistor
in a 50 Ohm Line

Whether series diodes or shunt diodes are used, the bias structures, dc returns and dc block structures are the same as those described in earlier chapters for mixers and other functions. It is also possible to incorporate the PIN diode as part of one of these structures, as, for example, on the end of a quarter- or half-wave stub. However, this construction tends to be narrower in bandwidth than either the series or the shunt diode when applied directly to the main transmission line. It is also important in diode switch construction that the bandwidth necessary for the switching speed of the diode be considered since structures must be used which will permit the necessary rise time characteristics to be achieved.

Multiple-Diode Construction

Since one diode is only capable of 15 to 20 dB of attenuation, circuit configurations making use of multiple diodes are generally necessary for high isolation switches. Because the diodes are reflective in their high isolation state, it is necessary to space these reflections properly in order to have additive isolation. This spacing, in general, should be approximately one quarter of a wavelength as shown in Figure 8–28.

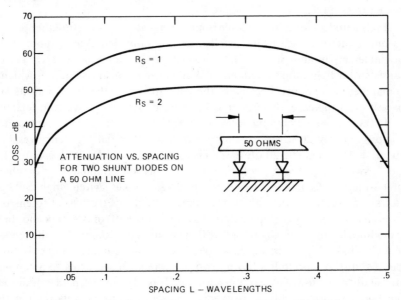

FIG. 8–28   Attenuation vs. Spacing for Two Shunt Diodes Mounted on a 50 Ohm Line

It can be seen from this figure, however, that the spacing is not critical. Therefore, minor modifications to diode spacing can be made along with line impedance shifts in order to minimize the mismatches of the diodes in the low attenuation state, thus improving low-loss conditions such as VSWR and insertion loss. Additional diodes may be added as isolation requirements increase, with their spacing again being approximately one quarter of a wavelength modified by matching requirements.

Modules

Because of package parasitic problems, many microwave engineers are now turning to integrated circuit modules for solid state control rather than going through the difficult task of using packaged diodes in their own circuit.[11] [12] The concept of solid state control modules is not new, but it has reached maturity in recent years with the development of improved PIN chips and the combination of improved packages and clever circuit design. A number of package styles have evolved, resulting from the electrical requirements of the modules and the manufacturers' preferences. These are illustrated in Figure 8–29.

In its simplest form, the module is merely a matching structure that permits the semiconductor device to be installed in a 50 ohm line. An example of this is the single-chip module which can be used either as an attenuator or as a switch. It consists of a PIN diode mounted in shunt to the transmission line, which is then matched to 50 ohms. Because of the low capacity of the diode and the lack of parasitics in the case, this device is broadband. Typically, over the region of 1.0 – 12.0 GHz, single-chip modules will offer less than 0.5 dB of insertion loss and approximately 20 dB of attenuation with a rather flat frequency response of about ± 1.0 dB at the 20 dB level. Therefore, they can be used as either low isolation switches or more frequently as low isolation attenuators. To achieve greater isolation, as in the case of the packaged diode switch, high isolation modules are usually made with two or more chips properly spaced to enhance the overall isolation. Two-chip modules having 40 to 45 dB of isolation and three-chip modules having typically 60 to 70 dB of isolation are common on the market. Like their single-chip brothers, these switches tend to be made using only shunt diodes. Multi-throw switches having two or more switched arms to a common input usually make use of at least one series diode per leg and then have additional shunt diodes to provide higher isolation. The series diode permits the junction

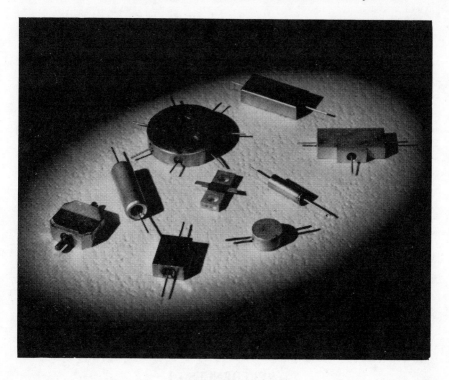

FIG. 8–29   Photograph of Several Module Styles for Stripline
Applications

to be matched without frequency sensitivity. As a result, this type of switch is available from UHF to 18.0 GHz in a single module.

The key to proper module use is installation in the package in such a fashion that the inherent isolation of the module structure is not degraded by the package. This installation method is shown in Figure 8–30 and illustrates the importance of the module shell.

Properly applied, the module represents the best possible switch configuration for low power applications in stripline, although it may still be necessary to use packaged diodes for high power requirements. As the cost of switching modules comes down, their use will undoubtedly become more universal and the problems of diode package parasitics will be eliminated.

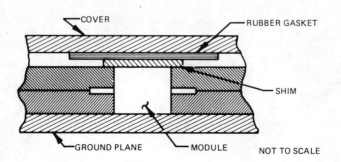

FIG. 8–30 Typical Module Installation

REFERENCES

1   Van Wagoner, R. C., *Multi-Octave Bandwidth Microwav Mixer Circuits,* MTT Symposium Digest 1968, pp. 8–15.

2   Howe, H., *Microwave Interferometers,* Micronotes, Vol. 9, No. 1, February, 1972.

3   Gerst, C., *Microwave Signal Processing Applications,* U. S. Patent Number 3,517,309.

4   Neuf, D., *A Doubly Balanced Mixer using Balanced Microstrip Baluns,* U. S. Patent Number 3,652,941.

5   Mouw, R. B., *A Broadband Hybrid Junction and Application to the Star Modulator,* MTT-16, No. 11, November, 1968, pp. 911–918.

6   Mouw, R. B., and Fukuchi, *Broadband Double Balanced Mixer/Modulators, Part I,* Microwave Journal, Vol.12, No. 3, March 1969, pp. 131; *Part II,* Microwave Journal Volume 12, No. 5, May 1969, p. 71.

7   *A Comparison of Various Doppler Mixer Diodes,* Micronotes, Volume 3, No. 6, October-November 1965.

8   Ashley, Lord, Mean, and Clarke, *Image Recovery Mixers,* Proceedings of the 1971 European Microwave Conference Vol. I, pp. A-11, 5:1–4, Stockholm, August 1971.

9   Aspesi, F. and Archangelo, T. D., *Low Noise Down-Converter for Radio Link Applications,* Proceedings of 1971 European Microwave Conference, Vol. I, pp. A-12/2:4, Stockholm, August, 1971.

10   Neuf, D., *A Quiet Mixer,* Microwave Journal, Vol. 16, No. 5, May, 1973, pp. 29–32.

11   Howe, H., *Modules for Solid-State Control,,* Microwave Journal, Vol. 15, No. 7, July 1972, pp. 14–17.

12   Reid, M. J., *Microwave Switch and Attenuator Modules,* Microwave Journal, Vol. 16, No. 7, July 1973, pp. 37–40.

—9—

Construction, Packaging, and Subassembly Techniques

The circuit configurations, design equations, and curves presented in the previous chapters are of no use unless the circuit functions developed from them can be mechanically packaged in a configuration which will perform reliably under whatever environmental conditions are required and which can be manufactured at a cost which is appropriate for the particular product. Most companies, after experimenting with various mechanical packaging techniques, have chosen one, two, or possibly three which suit their requirements better than others. As in the case of the materials, packaging techniques vary as a function of the product requirements and of the company's facilities to manufacture the product. A number of standard packaging techniques have been refined through the years and are now in common use.

1. Flat-Plate Construction

The simplest and perhaps the least expensive technique for building microwave stripline assemblies is the use of two flat plates on each side of the stripline board materials. This configuration is shown in Figure 9–1. The plates, which are generally clamped together by screws or rivets, may vary in thickness as long as they provide some edge mounting for the screws which hold the connectors onto the assembly. There are sealing problems associated with this type of construction, but it is ideal for simple breadboard circuit evaluation because it allows uniform pressure to be applied to the boards and provides good grounding for the connector flanges, thus simulating to a great extent, whatever further construction techniques may be used to generate the final product package.

293

FIG. 9–1   Flat Plate Construction

2. Bonded Metal Construction

A similar approach to the flat plate construction is construction wherein a reasonably thick material, usually aluminum, is bonded to the stripline dielectric material on one side, with a thin copper conductor bonded on the other side. After photo-etching, the pattern is left on one side of the board and the thick metal material is left on the other.

When sandwiched, an assembly similar to that shown in Figure 9–2 results. The end effect is very similar to the flat plate construction, although, in most cases, the metal material bonded to the dielectric is not thick enough to support edge threaded connector screws. Therefore, "U" channel connectors must be used for this type of assembly. While popular some years ago, this technique, like the previous one, is used largely for breadboard assemblies.

FIG. 9–2   Sandwich Construction using Heavy Bonded
Ground Planes

3. Channelled Chassis Construction

Either machined or stamped chassis can be used for channel construction. The machined type is shown in Figure 9–3. The actual pattern path of the conductors is machined in the chassis so that metal walls are maintained in reasonable proximity to both sides of the line. This technique is particularly useful when very high isolation circuits are required, or when circuit elements, such as diode switch modules or ferrite devices, are to be integrated into the stripline assembly. The major disadvantage of this approach is that the dielectric material must be machined in the form of "puzzle" pieces to fit the machined chassis channels. This, of course, can be expensive.

An alternate approach to the channelled chassis is a stamped or hydroformed chassis made of extremely thin sheet metal. This is shown in Figure 9–4.

A central web with double registration and air dielectric is suspended between two stamped sheet metal chassis which are then bonded together. The assembly is remarkably rigid and extremely strong, but unfortunately is useful only in environments when no sealing against humidity or water is required, although it is likely that this one drawback could be rapidly overcome should the need arise.

FIG. 9–3   Channel Chassis Construction

4. Box and Cover Construction

A deep covered box in which the stripline boards sit is also a traditional form of construction of stripline assemblies. Its configuration is shown in Figure 9–5.

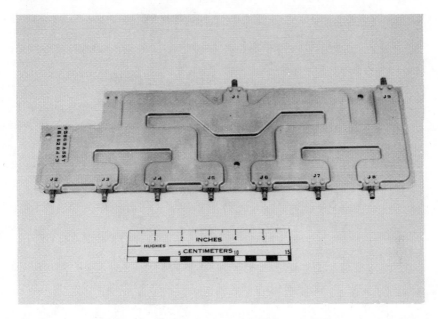

FIG. 9–4   Sheet Metal Channel Chassis Construction

It is relatively simple, straightforward, extremely sturdy, and resistant to environmental problems. However, there are difficulties in mounting active devices and in assembly techniques which will increase construction costs. A modified version of the box and cover technique uses a deeper box and the two stripline boards with a pressure plate on top of them. A sealing cover, which does not actually contact the internal assembly, completes the package. The connectors for this type of construction can be press-in, threaded-hole, or flange-type connectors, in contrast to the limitation on the split-block channelled chassis, which demands flange-type connectors.

5. No-Chassis Construction

What would appear to be the simplest and most straightforward approach, no chassis at all, is actually not simple. The technique of using no chassis, pioneered originally with Tri-plate. (Copyright Sanders Associates, Nashua, New Hampshire) components, uses just the dielectric boards with the copper ground plane on the outside and rivets to secure the entire assembly. A certain amount of rigidity is obtained, but there are sealing and strength problems. Then, too, the extremely thin ground plane is subject to damage which, if it occurs in the

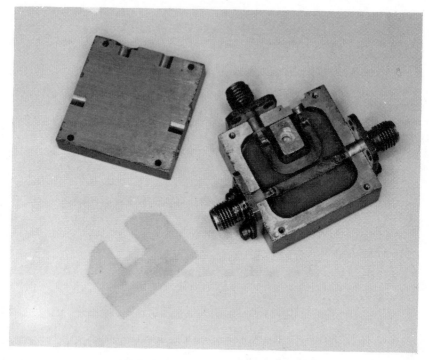

FIG. 9–5   Box-and-Cover Construction

proper area of the circuit, can cause electrical degradation. A modern version of this technique is shown in Figure 9–6 and is generally known as bonded stripline.

Either a thin dielectric material or adhesive material is inserted between the boards which are then combined under heat and pressure until they are fused into a solid assembly. Thus, no rivets or screws are necessary to hold the assembly together except as needed for mode suppression; for that application, plated-through holes may be substituted. This technique utilizes the lightest construction, but it does have drawbacks because it is totally irreparable, and does not lend itself to the integration of any type of active element which cannot stand the heat and pressure necessary to bond the assembly together. Nevertheless, this approach is receiving considerable attention and is being widely used in a number of lightweight, low-volume systems. It has the advantage of extreme environmental stability, particularly in the area of humidity and salt spray, since once the assembly is bonded together, it is impossible for the internal circuit to be attacked except through the connector interfaces, which still present a problem for many manufacturers.

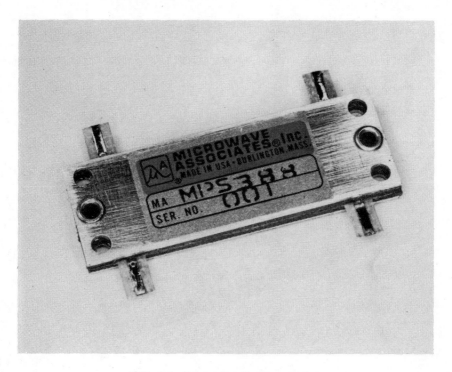

FIG. 9–6   Bonded Stripline Construction

In choosing a package construction, it is important that the user determine what the environment is likely to be, what the cost requirements are, and what circuit elements are to be contained in the package. Only when these factors are considered can an appropriate packaging technique be established. Certainly the worst thing which can be done is to establish the packaging technique before determining the package requirements.

Sealing

All of the packaging techniques previously discussed require that they be sealed against rain, humidity, salt spray, and in some cases, dust. A number of sealing techniques have evolved, many of which are applicable to one or more of the packaging methods. The simplest and least expensive material is epoxy, which can be used to seal the edges and seams or to encapsulate the total microwave assembly. This is illustrated in Figure 9–7. Like the bonded package, however, this technique has the disadvantage that it cannot be repaired once it is manufactured.

FIG. 9–7   Epoxy Encapsulated Stripline Assembly

Rubber and conductive-rubber gaskets are widely used in sealing machined chassis because gasket grooves can be readily machined. Alternately, flat gasket material can be placed underneath the cover prior to final assembly. If conductive gaskets are used, either technique serves a dual function of RFI protection and humidity and dust protection.

It is also possible to solder the assembly together using thin foil at the edges. Depending upon the circuit elements, however, this can be a tedious technique and is frequently difficult to implement. As a rule, soldered assemblies are not recommended, although they may be desirable in the case of hermetically sealed packages. If solder is considered as a sealing technique, the temperature of the solder should be chosen carefully. Low temperature solders will permit the sealing of the package, while still permitting disassembly of the package without destruction of the internal circuit elements which have also been soldered. Although useful for certain specific applications, solder has not gained general acceptance as a method of sealing stripline packages.

Metallized tape with adhesive backing has been used to seal flat-plate packages and other types of construction where the environmental requirements are not severe. It has a particular value for commercial applications because of its extremely low cost and its ability to be removed if necessary. This type of tape is available with both conductive and non-conductive adhesives. When properly applied, a good quality tape with a thermo-setting adhesive which can be baked after application will provide a very high degree of package integrity. It is by nature thin, and any mechanical abrasion can readily break through it or tear it from the edge, thereby destroying the package integrity. Nevertheless, as tape materials and tape adhesives improve, the application of taped package sealing will increase.

FIG. 9–8   TIG Welding Applied to the Chassis Edge

When the stripline package contains a non-hermetically sealed semiconductor device, or when environmental conditions are extreme, it may be necessary to hermetically seal the entire package. This requires the use of glass bead connectors which are soldered in, thus maintaining the hermeticity at the connector interfaces. The overall assembly must be sealed, evacuated, and then backfilled with an inert gas.

Although solder techniques operate at lower temperatures, there is growing evidence that for a truly hermetically sealed assembly, welding represents the best approach to the problem. Techniques have been developed primarily for ceramic microstrip packages making use of aluminum chassis and TIG (Tungsten Inert Gas) type welding*. This type of weld fuses the edges of a thin section of the chassis so that, if the chassis is properly heat sunk, the overall chassis can be welded shut without raising its internal temperature beyond the limits of the normal dielectric materials and semiconductor devices which may be within. Figure 9–8 illustrates TIG welding of a test chassis and shows the welding head travelling along the edge of the chassis while welding it closed.

One significant factor of the welding approach is that, although it is permanent, it is possible to take a light skim cut off the top of the package and remove the cover for repair. With proper package design, this can be done two or three times before the basic chassis becomes too thin to permit rewelding.

Circuit Layout

The ultimate goal of the stripline package is to integrate many components, for there are many other techniques of building single microwave components which may be superior to stripline. The virtue of stripline lies in its ability to integrate a number of components and subassemblies due to its two-dimensional nature and planar construction. In doing this, there are certain ground rules and techniques which can be applied in order to increase the probability of success, most important of which are the methods of layout. Circuit elements should be laid out so that they do not tend to interact with each other. In the case of parallel lines, this means that a minimum line-spacing of two strip-widths should be maintained between parallel lines when coupling is not desired. Similarly, it is desirable to eliminate as many bends as possible. Although wide-radii, and mitered bends have minimum VSWR contribution, a multitude of these bends can introduce so many uncontrolled reflections that the overall circuit performance may not be achieved. It is also important to lay out the individual circuit functions in a logical flow, within the restraints of the available package dimensions and possible customer-determined connector locations. Whenever

*U. S. Army Electronics Command, Fort Monmouth, New Jersey, Contract Number DAAB07–73-C-0068.

possible, high signal level, and low signal level circuits should be placed on opposite sides of the package so that adequate isolation can be maintained between them. An example of logical flow within the constraints of the package area available for construction is shown in Figure 9–9.

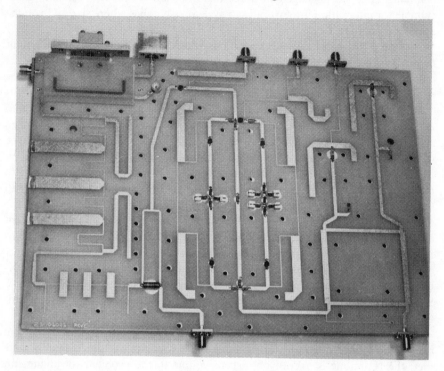

FIG. 9–9 Multi-Component Package Circuit Layout

An input directional coupler feeds a five-section bandpass filter, which feeds a low-pass filter used for harmonic suppression and a power combiner whose function is to inject a test signal. The output of this combiner then feeds a SP3T diode switch, the three legs of which contain 0, 20, and 40 dB step attenuators. These three output legs are then recombined in a SP3T output diode switch, which feeds a balanced mixer. The signal flow through the package is a meander making optimum use of the available area for circuit construction. This meander approach may also be used on a smaller scale to compress the dimensions of low frequency circuits. At frequency ranges in the order of 1.0 GHz or below, the quarter-wave sections, needed for many of the branch type structures described, become excessively long. There is no reason why they must be run in a straight line. Meandering can be used to fill the area, and thus condense the overall size of the circuit function at these frequencies. Each bend in the meander should be matched by one of the techniques described in Chapter 2 in order to minimize internal VSWR's in the circuit element. These techniques, in combination with some of the higher dielectric materials, permit low frequency components to be packaged in a reasonably small space, thus making them suitable for integration into overall sub-system assemblies where size and volume are a constraining factor.

One of the most difficult conditions which can be imposed on the circuit layout designer is the pre-determined location of connectors. This means that the circuit layout must be modified to conform to the input and output locations which may not be determined by logical circuit functions. Inasmuch as this condition is usually dictated by the customer, the condition must be tolerated and the designer must be clever enough to work around the limitations. When such conditions exist, most of the rules concerning logical flow and the avoidance of bends must, of necessity, be ignored, and performance is likely to suffer.

In extreme cases of mechanical constraint, it may be necessary to go to multi-layer construction using broad-wall feedthroughs similar to the broad-wall launchers described in Chapter 2. This technique is effective for compressing the dimensions of an overall package and is usable at frequency ranges up to the point at which the broad-wall interface connectors are no longer satisfactory. This point is a function not only of frequency but of bandwidth. For most cases, multi-layer construction is not recommended at frequencies above high C-band or low X-band. An example of multi-layer construction in C-band is shown in Figures 9–10a and 9–10b, which represent the upper and lower decks of a two layer multi-channel C-band front end.

Comparison of the photographs will show the various feedthroughs from deck to deck. Although all technical parameters required of this assembly were met, the technique is excessively expensive due to the problems of alignment, registration, and feedthrough.

Circuit Isolation

Perhaps one of the most significant problems when integrating many circuit elements in the same package is the isolation between circuits, or the isolation across any given circuit. It is generally impossible to achieve more than 70 dB of reliable isolation between any two unconnected circuit elements in a common

FIG. 9–10 Upper and Lower Decks of a Multi-Channel Two-Layer C-band Front-End

set of ground planes. This number varies slightly with the distance between the circuit elements and the frequency of operation. Thus, in order to have circuits operate with high degrees of isolation, it is necessary to add some type of mode suppression in order to prevent the parallel-plate modes, which are launched due to minor circuit discontinuities, from decreasing the achievable isolation. One of the most simple and effective techniques is the use of screws to bring the two surfaces of the parallel-plate construction to the same potential. In order to be effective, it is necessary that these screws be placed less than 1/8 of a wavelength from each other at the frequency of concern. In the case of high frequency circuits, this can result in a large number of screws, eyelets, rivets, or whatever is used to generate the mode suppression. Where flat-plate or no-chassis type construction is used, thin walls, which can also be inserted by means of foil, shims, or other techniques, can be employed to increase isolation. However, the cost of these techniques is frequently greater than what would be spent in building channelled construction. Thus, unless the user is committed to this type of fabrication for other overriding needs, it should be avoided.

Active Element Integration

Many circuits require the use of active elements. These may be shunt- or series-mounted and, depending upon their function, i.e., PIN switch diodes or mixer diodes, it may be desirable to have them be replaceable. In the case of replaceable semiconductor devices, the most useful technique is to use a shunt configuration which permits edge-mounted diode holders. A typical diode holder was shown in Figure 8–10.

When a shunt-mount diode is not practical, access hatches which will permit replacement of series diodes may be employed. It is important, however, in the use that access hatches, to be certain that the ground plane around the semiconductor element is continuous. Therefore, a multi-layer access hatch is generally required.

Since the reliability of microwave semiconductors has increased in recent years, non-replaceable diodes have become a more desirable approach, in terms of simplicity, lower cost, and long term reliability, because replaceable diodes tend to be replaced frequently even though they are not actually defective. Non-replaceable diodes may be installed by means of either the modules or glass axial–lead solder-in diodes. Either of these techniques is perfectly acceptable, and the approach is determined by frequency response, bandwidth, and the type of semiconductor device to be used.

The single most important concept in microwave packaging is to include within one package all functions necessary to provide a cohesive unit, rather than to break the components down into smaller packages which may be limited merely by the capability of the manufacturers. Any component that is logically and functionally desirable should be constructed within the package, regardless of the transmission line technique necessary.

Stripline is the most effective base technology for integration, but is not necessarily the only one applicable. It may also be necessary to integrate both low-frequency circuits in order to provide driver or preamplifier functions, and waveguide, coaxial lines, or ceramic microstrip in order to provide additional microwave functions.

Because no firm rule governs the amount of integration needed or the exact packaging technique to be employed, the level of integration is a major factor after the decision to integrate a subassembly has been made. Under normal circumstances, the integration of components into a common assembly results in reduced expenses due to lower material costs, elimination of connectors, and more efficient testing.

FIG. 9–11   Multi-Media Ku-band Assembly

These benefits can be overshadowed by financial losses due to over-integration of the package, because critical circuits or unproven components can result in poor production yields. When critical circuits are necessary, they should be designed in a modular form so that they can be pretested and inserted as individual subcomponents. Typical of this technique is the attachment of the waveguide filter shown in Figure 9–11, which is a Ku-band assembly containing a number of directional couplers and ferrite junctions, a balanced mixer, an IF preamplifier, and a high-Q waveguide filter. In addition, the system output antenna and two waveguide Gunn sources are mounted on the assembly. Thus, direct transitions have been made from waveguide to stripline, and back to waveguide.

Instead of basing the package construction upon the most critical single component or upon the available technology, it may be desirable to mix the packaging techniques within a common package. In Figure 9–13, inserting a waveguide filter within the stripline package results in lower losses than those achievable with pure stripline construction. Then, too, a combination of stripline and ceramic microstrip is sometimes appropriate in order to make use of the extremely broad bandwidth capabilities of stripline, mixed with the non-parasitic semiconductor mounting techniques achievable with ceramic microstrip. This mixed-media approach can permit each component to function at maximum efficiency, and if the transitions are made in direct and logical fashion, the integrity of the package can be maintained.

Bibliography

Abeyta, I., *Design Techniques for Strip Line Directional Couplers and Resonators,* Army Signal Engineering Laboratory (Fort Monmouth, New Jersey), June 10, 1957, AD146739, pp. 633–638.

Adair, J. E., and Haddad, G. I., *Coupled-Mode Analysis of Non-Uniform Coupled Transmission Lines,* MTT-17, No. 10, October 1969, pp. 746–752.

Adams, D. K., and Weir, W. B., *Wideband Multiplexers Using Directional Filters,* Microwaves, Vol. 8, No. 5, May 1969, pp. 44–50.

Albanese, V. J., and Kagen, H., *The "Cross-Over" Directional Coupler,* Microwave Journal, Vol. 4, No. 9, September 1961, pp. 88–91.

Albanese, V. J., and Peyser, W. P., *An Analysis of a Broad Band Coaxial Hybrid Ring,* MTT-6, No. 4, October 1958, pp. 367–373.

Alford, A., *Coupled Networks in Radio-Frequency Circuits,* IRE Proceedings, Vol. 29, February 1941, pp. 55–70.

Alford, A., and Watts, C. B., *A Wide-Band Coaxial Hybrid,* IRE Convention Record, 1956, Part 1, pp. 176–179.

Allen, J. L., and Estes, M. F., *Broadside Coupled Strips in a Layered Dielectric Medium,* MTT-20, No. 10, October 1972, pp. 662–668.

309

Alstadter, D., and Houseman, E. O., Jr., *Some Notes on Strip Transmission Line and Waveguide Multiplexers,* IRE Wescon, December 1958, Part 1, pp. 54–69.

Alstadter, D., and Houseman, E. O., Jr., *Strip Transmission Line Corporate Feed Structures for Antenna Arrays,* IRE National Convention Record, 1959, Vol. III, Part 3, pp. 113–125.

Altschuler, H. M., and Oliner, A. A., *Discontinuities in the Center Conductor of Symmetric-Strip Transmission Line,* MTT-8, No. 3, May 1969, pp. 328–339.

Anderson, J. C., *The Calculation of Characteristic Impedance by Conformal Transformation,* Journal of the British IRE, Vol. 18, January 1958, pp. 49–54.

Arndt, F., *Tables For Asymmetric Chebyshev High Pass TEM Mode Directional Couplers,* MTT-18, No. 9, September 1970, pp. 633–638.

Badoyannis, G. M., *The Power Handling Capacity of Striplines,* IRE Convention Record, 1958, Part 1, pp. 35–38.

Barker, L. R., *Strip Line Coupler Design Chart,* Microwaves, March 1967, p. 46.

Barnett, E., Lacy, P., and Oliver, B., *Principle of Directional Coupling in Reciprocal Systems,* Proceedings of the Symposium on Modern Advances in Microwave Technology, Polytechnic Institute of Brooklyn, Vol. 14, November 1954, pp. 383–392.

Barrett, R., *Etched Sheets Serve As Microwave Components,* Electronics, Vol. 25, June 1952, pp. 114–118.

Barrett, R., *Microwave Printed Circuits — A Historical Survey,* MTT-3, No. 2, March 1955, pp. 1–9.

Barrow, C. R., *The Exponential Transmission Line,* Bell Systems Technical Journal, Vol. 17, October 1938, pp. 555–573.

Begovich, N., *Capacity and Characteristic Impedance of Strip Transmission Lines with Rectangular Inner Conductors,* MTT-3, No. 2, March 1955, pp. 127–133.

Begovich, N., and Margolin, A. R., *Theoretical and Experimental Studies of a Strip Transmission Line,* Hughes Aircraft Company Technical Memo 234, May 12, 1950.

Bogner, B. F., *Ultra Lightweight Stripline RF Manifold,* Proceedings of the Eleventh Electrical Insulation Conference, 1973, pp. 278–281.

Bolinder, E. F., *Fourier Transformers and Tapered Transmission Lines,* IRE Proceedings, Vol. 44, April 1956, p. 557.

Bostick, G., *Design Procedure for Coaxial High-Pass Filters,* Electronics Design, No. 8, April 1969, pp. 66–69.

Bouthinon, M., and Coumes, A., *Broadband Hybrids,* MTT-15, No. 7, July 1967, pp. 431–432.

Bowen, E. E., *Forward with Backward Diodes,* Electronic Design, April 26, 1966, pp. 44–49.

Boyd, C. R., Jr., *On a Class of Multiple-Line Directional Couplers,* MTT-10, No. 14, July 1962, pp. 287–294.

Bradley, E. H., *Design and Development of Stripline Filters,* MTT-4, No. 2, April 1956, pp. 86–93.

Bradley, E. H., and White, D. R., *Bandpass Filters Using Strip-Line Techniques,* Electronics, May 1955, pp. 152–155.

Bradley, E. H., and White, D. R., *Bandpass Filters Using Strip Line Techniques,* MTT-3, No. 2, March 1955, pp. 163–169.

Breithaupt, R. W., *Conductance Data for Offset Series Slots in Stripline,* MTT-16, No. 1, November 1968, pp. 969–970.

Brown, N. J., and Basken, P., *Design Concepts for IFF Diode Switches,* Microwave Journal, Vol. 16, No. 7, July 1973, pp. 43–48.

Browness, C., *Strip Transmission Lines,* Electronic Engineering, Vol. 28, January 1956, pp. 2–7.

Budenbon, H. T., *Transmission Properties of Hybrid Rings and Related Annuli,* IRE National Convention Record, 1957, Vol. 5, Part I, p. 186.

Bunker, J., and Howe, H., *Stripline/Microstrip, A Mixed Media Approach for Advanced ECM Systems,* Conference Proceedings of Microwave '73 (Brighton, England), June 1973, pp. 143–148.

Campbell, J. J., *Application of the Solutions of Certain Boundary Value Problems to the Symmetrical Four-Port Junction and Specially Truncated Bends in Parallel Plate Waveguides and Balanced Strip Transmission Lines,* MTT-16, No. 3, March 1968, pp. 165–176.

Cappucci, J., *Don't Overspecify with Quad Hybrids, Part I,* Microwaves, Vol. 12, No. 1, January 1973, pp. 50–54. *Part II,* Microwaves, Vol. 12, No. 2, February 1973, pp. 62–66.

Cappucci, J., *Zero-Loss Wide-Band Coaxial Line Variable Attenuators,* Microwave Journal, Vol. 3, No. 6, June 1960, pp. 49–53.

Carlin, H. J., *Cascaded Transmission-Line Synthesis,* Polytechnic Institute of Brooklyn, Report P1BMR1 889–61, April 1961.

Carlin, H. J., *A UHF Multiplexing System Using Frequency Selective Coaxial Directional Couplers,* IRE Convention Record, Vol. III, 1955, p. 135.

Carpenter, E., *An Asymmetric Non-Monotonic Stripline Magic Tee,* MTT Symposium Digest, 1969, pp. 320–323.

Carpenter, E., *The Virtues of Mixing Tandem and Cascade Coupler Connections,* MTT Symposium Digest, 1971, pp. 8–9.

Carr, J. W., *Balanced Line UHF and L-Band Bi-Symmetrical Hybrids,* Microwave Journal, Vol. 7, No. 9, September 1964, pp. 47–50.

Carr, K., and Howe, H., *Let's Remove the Mystery from Microwave Packaging,* Microwaves, March 1973, Vol. 12, No. 3, pp. 42–47.

Carson, C. T., and Cambrell, G. K., *Upper and Lower Bounds on the Characteristic Impedance of TEM Mode Transmission Lines,* MTT-14, No. 10, October 1966, pp. 497–498.

Cetaruk, W., *Computer Draws Art Work for Strip-Line,* Microwaves, November 1966, pp. 4852.

Chao, G., *A Wide-Band Variable Microwave Coupler,* MTT-18, No. 9, September 1970, pp. 576–583.

Coale, F. S., *Applications of Directional Filters for Multiplexing Systems,* MTT-6, No. 4, October 1958, pp. 450–453.

Coale, F. S., *Strip Line Excitation Methods,* Radio Television News, January 1955.

Coale, F. S., *Strip Lines,* Sperry Engineering Review, No. 8, July/August 1955, pp. 25–29.

Coale, F. S., *A Traveling Wave Directional Filter,* MTT-4, No. 4, October 1956, pp. 256–260.

Cohen, J., *Design of a Stripline Wideband Hybrid Ring,* Microwave Journal, Vol. 13, No. 8, August 1970, pp. 32a – 32e.

Cohn, S. B., *Parallel-Coupled Transmission-Line Resonator Filters,* MTT-6, No. 2, April 1958, pp. 223–231.

Cohn, S. B., *Characteristic Impedance of the Shielded-Strip Transmision Line,* MTT-2, No. 2, July 1954, pp. 52–57.

Cohn, S. B., *A Class of Broadband Three-Port TEM-Mode Hybrids,* MTT-16, No. 2, February 1968, pp. 110–118.

Cohn, S. B., "Design of Transmission-Line Filters", Chapter 27 of *Very High Frequency Techniques,* Vol. 1, H. J. Reich, Editor, McGraw-Hill (New York), 1947.

Cohn, S. B., *Direct-Coupled Resonator Filters,* IRE Proceedings, Vol. 45, February 1957, pp. 187–196.
Correction to "Direct-Coupled-Resonator Filters", IRE Proceedings, Vol. 45, No. 7, July 1957, p. 956.

Cohn, S. B., *Dissipation Loss in Multiple-Coupled Resonator Filters,* IRE Proceedings, Vol. 47, August 1959, pp. 1342–1348.

Cohn, S. B., *New Thoughts on Stripline,* Microwave Journal, Volume 5, No. 7, July 1962, pp. 13–18.

Cohn, S. B., *Optimum Design of Stepped Transmission Line Transformers,* MTT-3, No. 3, April 1955, pp. 16–21.

Cohn, S. B., *Characteristic Impedance of Broadside-Coupled Strip Transmission Lines,* MTT-8, No. 6, November 1969, pp. 633–637.

Cohn, S. B., *Problems in Strip Transmission Lines,* MTT-3, No. 2, March 1955, pp. 119–126.

Cohn, S. B., *A Re-appraisal of Strip-Transmission Line,* Microwave Journal, Volume 3, No. 3, March 1960, pp. 17–27.

Cohn, S. B., *Shielded Coupled-Strip Transmission Line,* MTT-3, No. 5, October 1955, pp. 29–37.

Cohn, S. B., *Thickness Corrections for Capacitive Obstacles and Strip Conductors,* MTT-8, No. 6, November 1960, pp. 638–644.

Cohn, S. B., et al, *Strip Transmission Lines and Components,* Final Report SRI Project 1114, Contract DA36–039-SC-63232, DA Project 3–26–00–600, Stanford Research Institute, March 1957.

Cohn, S. B., and Coale, F. S., *Directional Channel-Separation Filters,* IRE Proceedings, 1956, No. 44, p. 1018.

Cohn, S. B., and Koontz, R. H., *Microwave Hybrid Coupler Study Program,* Third Quarterly Progress Report, Contract DA36–239-SC-87435, Rantec Corporation, January 1962.

Cohn, S. B., and Sheck, P. M., et al, *Strip Transmission Lines and Components,* Antenna Systems Laboratory, Stanford Research Institute, Final Report, February 1957, AD145251.

Collin, R. E., *The Optimum Tapered Transmission-Line Matching Section,* IRE Proceedings, Vol. 44, No. 4, April 1956, pp. 539–548. *Correction to "The Optimum Tapered Transmission-Line Matching Section",* IRE Proceedings, Vol. 44, No. 12, December 1956, p. 1753.

Collin, R. E., *Theory and Design of Wide-Band Multi-Section Quarter-Wave Transformers,* IRE Proceedings, Vol. 43, February 1955, pp. 179–185.

Coltum, L., and Torgow, E. N., *Delay Characteristics of Basic Strip Transmission Line Structure,* Microwave Research Institute, Polytechnic Institute of Brooklyn, March 13, 1956, AD99013.

Cristal, E. G., *Analysis and Exact Synthesis of Cascaded Commensurate Transmission-Line C-Section All-Pass Networks,* MTT-14, No. 6, June 1966, pp. 285–291.

Cristal, E. G., *Band-Pass Spur Line Resonators,* MTT-14, No. 6, June 1966, pp. 296–297. Correction to *"Band-Pass Spurline Resonators",* MTT-14, No. 9, September 1966, p. 436.

Cristal, E. G., *Coupled-Transmission-Line Directional Couplers with Coupled Lines of Unequal Characteristic Impedances,* MTT-14, No. 7, July 1966, pp. 337–346.

Cristal, E. G., *Design Equations for a Class of Wide-Band Bandpass Filters,* MTT-20, No. 10, October 1972, pp. 696–699.

Cristal, E. G., *Meander-Line and Hybrid Meander-Line Transformers,* MTT-21, No. 2, February 1973, pp. 69–75.

Cristal, E. G., *New Design Equations for a Class of Microwave Filters,* MTT-19, No. 5, May 1971, pp. 486–490.

Cristal, E. G., *Nonsymmetrical Coupled Lines of Re-entrant Cross Section,* MTT-15, No. 9, September 1967, pp. 529–530.

Cristal, E. G., *Re-Entrant Directional Couplers Having Direct Coupled Center Conductors,* MTT-14, No. 4, April 1966, pp. 207–208.

Cristal, E. G., *Tables of Maximally Flat Impedance-Transforming Networks of Low-Pass Filter Form,* MTT-13, No. 5, September 1965, pp. 693–695.

Cristal, E. G., and Frankel, S., *Hairpin-Line and Hybrid Hairpin-Line/Half-Wave Parallel-Coupled-Line Filters,* MTT-20, No. 11, November 1972, pp. 719–728.

Cristal, E. G., and Gysel, U. H., *New Compact Multiplexer for Stripline and MIC,* Proceedings of the European Microwave Conference, 1973, Vol. 1, p. B.3.1.

Cristal, E. G., and Matthaei, G. L., *A Technique for the Design of Multiplexers Having Contiguous Channels,* MTT-12, No. 1, January 1964, pp. 88–93.

Cristal, E. G., and Young, L., *Theory and Tables of Optimum Symmetrical TEM-Mode Coupled Transmission-Line Directional Couplers,* MTT-13, No. 5, September 1965, pp. 544–558.

Croven, J. H., *Coaxial to Strip-Transmission Line Adaptor,* MTT-9, No. 2, March 1961, pp. 200–201.

Dahlman, B. A., *A Double Ground-Plane Strip-Line System for Microwaves* MTT-3, No. 5, October 1955, pp. 52–56.

Dalley, J. E., *A Stripline Directional Coupler Utilizing a Non-Homogeneous Dielectric Medium,* MTT Symposium Digest, 1967, pp. 63–65.

David, S., *A Wide-band Coaxial Line Power Divider,* MTT-15, No. 4, April 1967, pp. 270–271.

David, S., and Kahn, W. K., *Optimal 3-Port Power Dividers Derived from Hybrid Tee Prototypes,* MTT Symposium Digest, 1967, pp. 54–57.

Davis, R. M., *Three Quarter-Wave Parallel-Staggered Microwave Filters,* MTT-17, No. 7, July 1969, pp. 404–405.

DeBuda, R., *A Method of Calculating the Characteristic Impedance of a Strip Transmission Line to a Given Degree of Accuracy,* MTT-6, No. 4, October 1958, pp. 440–446.

Dent, J. R., *Stripline Technique Produces a Simple 3-dB Directional Coupler,* Electronic Design, August 31, 1960, pp. 52–53.

Drummer, G. W., and Johnston, D. L., *Printed and Potted Electonic Circuits,* IRE Proceedings, Paper No. 1407R, November 1952, 100, Part III, p. 178.

DuHamel, R. H., and Armstrong, M. E., *A Wide-Band Monopulse Antenna Utilizing the Tapered Line Magic-T,* Fifteenth Annual Symposium, AFAL, Wright-Patterson AFB, 1965.

Dukes, J. M. C., *The Application of Printed-Circuit Techniques to the Design of Microwave Components,* IRE Proceedings, Paper No. 2401R, August 1957, 105B, p. 155.

Dukes, J. M. C., *An Investigation into Some Fundamental Properties of Strip Transmission Lines with the Aid of an Electrolytic Tank,* IRE Proceedings, Paper No. 1991R, May 1956, 103B, p. 319.

Dukes, J. M. C., *Re-Entrant Transmission Line Filter Using Printed Conductors,* IRE Proceedings, Paper No. 244R, November 1957, 105B, p. 173.

Duncan, J. W., *Characteristic Impedances of Multi-Conductor Strip Transmission Lines,* MTT-13, No. 1, January 1965, pp. 107–118.

Earle, M. A., and Benedik, P., *Characteristic Impedance of Dielectric Supported Strip Transmission Line,* MTT-16, No. 10, October 1956, pp. 884–885.

Ekinge, R., and Hedstrom, T., *A New Variable Microwave Attenuator,* MTT-18, No. 9, September 1970, pp. 661–662.

Ekinge, R., *A New Method of Synthesizing Matched Broad-Band TEM-Mode Three-Ports,* MTT-19, No. 1, January 1971, pp. 81–88.

Ekinge, R., *Microwave TEM-Mode Power Dividers and Couplers,* Technical Report #15, Chalmers University of Technology (Goteborg, Sweden), 1972.

Ekinge, R., *On the Design of Impedance-Transforming Directional Couplers,* MTT-19, No. 4, April 1971, pp. 415–416.

Firestone, W. L., *Analysis of Transmission-Line Directional Couplers,* IRE Proceedings, Vol. 42, October 1954, pp. 1529–1538.

Fleri, D. and Hanley, G., *Nonreciprocity in Dielectric Loaded TEM Mode Transmission Lines,* MTT-7, No. 1, January 1959, pp. 23–27.

Foster, K., *The Characteristic Impedance and Phase Velocity of High-Q Tri-Plate Line,* Journal of the British IRE, Vol. 18, December 1958, pp. 715–723.

Franco, A. G., and Oliner, A. A., *Symmetric Strip Transmission Line Tee Junction,* MTT-10, No. 2, March 1962, pp. 118–124.

Fromm, W. E., *Characteristics and Some Applications of Stripline Components,* MTT-3, No. 2, March 1955, pp. 13–20.

Fromm, W. E., Fubini, E. G., and Keen, H. S., *A New Microwave Rotary Joint,* IRE National Convention Record, Vol. VI, 1958, Part 1, p. 78.

Frost, A. D., and Mingins, C. R., *Microwave Strip Circuit Research at Tufts College,* MTT-3, No. 2, March 1955, pp. 10–12.

Fubini, E. G., Fromm, W. E., and Keen, H. S., *New Techniques for High-Q Strip Microwave Components,* IRE Convention Record (8), No. 2, 1954, pp. 91–97.

Fubini, E. G., McDonough, J. A., and Malech, R., *Stripline Radiators,* IRE Convention Record, Vol. III, 1955, p. 51.

Fubini, E. G., *Stripline Radiators,* MTT-3, No. 2, March 1955, pp. 149–156.

Fubini, E. G., Fromm, W. E., and Keen, H. S., *Microwave Applications of High Q Strip Components,* IRE Convention Record (8), No. 2, 1954, pp. 98–103.

Galin, I., and Grayzel, A., *A New Type of Compact Wideband Stripline Filter*, Proceedings of the European Microwave Conference, 1973, Vol. 1, p. B.3.3.

Garault, Y., and Besse, M., *Ultra-Broadband Microwave Coupler Using Non-Uniform Striplines*, Proceedings of the European Microwave Conference, 1971, p. B12/6.

Gardiol, F. E., *Power Combiner Nomograms*, MTT-18, No. 1, January 1970, pp. 71–72.

Geppert, D. V., and Koontz, R. H., *TEM Mode Microwave Filters*, Tek-Tech and Electronics Industries, November 1955, pp. 72–73, 150–151.

Gerst, C. W., *Electrically Short 90° Couplers Utilizing Lumped Capacitors*, MTT Symposium Digest, 1967, pp. 58–62.

Getsinger, W. J., *Coupled Rectangular Bars Between Parallel Plates*, MTT-10, No. 1, January 1962, pp. 65–72.

Getsinger, W. J., *A Coupled Strip-Line Configuration Using Printed Circuit Construction That Allows Very Close Coupling*, MTT-9, No. 6, November 1961, pp. 535–544.

Gish, D. C., and Graham, O., *Characteristic Impedance and Phase Velocity of a Dielectrically-Supported Air Strip Transmission Line*, MTT-18, No. 3, March 1970, pp. 131–148.

Goodman, P. C., *A Wideband Stripline Matched Power Divider*, MTT Symposium Digest, 1968, pp. 16–20.

Green, H. E., *The Numerical Solution of Some Important Transmission Line Problems*, MTT-13, No. 5, September 1965, pp. 676–692.

Green, H. E., and Pyle, J. R., *The Characteristic Impedance and Velocity Ratio of Dielectric Supported Strip Line*, MTT-13, No. 1, January 1965, pp. 135–136.

Guckel, H., *Characteristic Impedances of Generalized Rectangular Transmission Lines*, MTT-13, No. 3, March 1965, pp. 270–274.

Gunderson, L. C., and Guida, A., *Stripline Coupler Design,* Microwave Journal, Volume 8, No. 6, June 1956, pp. 97–101.

Gupta, O. P., and Wenzel, R. J., *Design Tables for a Class of Optimum Microwave Bandstop Filters,* MTT-18, No. 7, July 1970, pp. 402–404.

Gupta, R. R., *Accurate Impedance Determination of Coupled TEM Conductors,* MTT-17, No. 8, August 1969, pp. 479–488.

Gysel, U. H., *Improved Hairpin Line Filters,* 1973 MTT Symposium Digest, pp. 205–207.

Hall, A. M., *Impedance Matching by Tapered or Stepped Transmission Lines,* Microwave Journal, Vol. 9, No. 3, March 1966, pp. 109–114.

Hallford, B., *A 90-dB Microstrip Switch on a Plastic Substrate,* MTT-19, No. 7, July 1971, pp. 654–657.

Hallford, B., *Low Noise Microstrip Mixer on a Plastic Substrate,* MTT Symposium Digest 1970, pp. 206–211.

Hamid, M. A., and Yunik, M. M., *On the Design of Stepped Transmission Line Transformers,* MTT-15, No. 9, September 1967, pp. 528–529.

Hartman, J. A., and Palermo, C. A., *Microwave Stripline Packaging Using Conductive Epoxy,* Proceedings of the 11th Electrical Insulation Conference, 1973, pp. 286–289.

Harvey, A. F., *Parallel-Plate Transmission Systems for Microwave Frequencies,* Proceedings IRE, Vol. 106, Part B, March 1959, pp. 129–140.

Havens, R. C., *An X-Band Strip-Transmission Line Tunnel Diode Amplifier,* Microwave Journal, Vol. 9, No. 5, May 1966, pp. 49–54.

Hinden, H. J., *Input VSWR and Output Isolation of Lossy N-Way Hybrid Power Dividers,* MTT-16, No. 3, March 1968, pp. 199–201.

Hinden, H. J., *Standing-Wave Ratio of Binary TEM Power Dividers,* MTT-16, No. 2, February 1968, pp. 123–125.

Hinden, H. J., and Rosenzweig, A., *3.0 dB Couplers Constructed from Two Tandem Connected 8.34 dB Asymmetric Couplers,* MTT-16, No. 2, February 1968, pp. 125–126.

Hinden, H. J., and Taub, J. J., *Design of TEM Equal Stub Admittance Filters,* MTT-15, No. 9, September 1967, pp. 525–528.

Horgan, J. D., *Coupled Strip Transmission Lines with Rectangular Inner Conductors,* MTT-5, No. 2, April 1957, pp. 92–99.

Horton, M. C., and Wenzel, R. J., *General Theory and Design of Optimum Quarter-Wave TEM Filters,* MTT-13, No. 3, March 1965, pp. 316–317.

Howe, H., *Broadband Stripline Directional Couplers,* Micronotes, Vol. 5, No. 7, May 1968.

Howe, H., *Broadband Stripline Mixers,* Micronotes, Vol. 6, No. 4, January 1969.

Howe, H., *Dielectrically Loaded Stripline at 18.0 GHz,* Microwave Journal, Vol. 9, No. 1, January 1966, pp. 52–54.

Howe, H., *Image Suppression Mixers and Single Sideband Modulators,* LEL Div. Varian Associates, Technical Memo, TM-0006.

Howe, H., *Interactive Time Sharing Techniques for TEM Mode Circuit Designs,* Conference Proceedings of Microwave '73 (Brighton, England), 1973, pp. 337–341.

Howe, H., *Microwave Interferometers,* Micronotes, Vol. 9, No. 1, February 1972.

Howe, H., *Modules for Solid State Control,* Microwave Journal, Vol. 15, No. 7, July 1972, pp. 14–17.

Howe, H., *Stripline is Alive and Well . . . .,* Microwave Journal, Vol. 14, No. 7, July 1971, pp. 25–28.

Howe, H., *Why Not Use Stripline?,* IEEE Conference Proceedings, New York, March 1972, pp. 488–489.

Hoyt, W. H., Jr., *Potential Solution of a Homogeneous Stripline of Finite Width,* MTT-3, No. 4, July 1955, pp. 16–17.

Jones, E. M. T., *A Wide-Band, Strip-Line Magic Tee,* MTT-8, No. 2, March 1960, pp. 160–168.

Jones, E. M. T., and Bolljahn, J. T., *Coupled Strip-Transmission-Line Filters and Directional Couplers,* MTT-4, No. 2, April 1956, pp. 75–80.

Jones, E. M. T., and Shimizu, J. K., *A Wide-Band Strip-Line Balun,* MTT-7, No. 1, January 1956, pp. 128–134.

Kaiser, J. A., *Ring Network Filter,* MTT-9, No. 4, July 1961, pp. 359–360.

Kammler, D. W., *Calculation of Characteristic Admittances and Coupling Coefficients for Strip Transmission Lines,* MTT-16, No. 11, November 1968, pp. 925–937.

Kammler, D. W., *The Design of Discrete N-Section and Continuously Tapered Symmetrical Microwave TEM Directional Couplers,* MTT-17, No. 8, August 1969, pp. 577–590.

Karakach, J. T., *Transmission Lines and Filter Networks,* McMillan, New York, 1950.

Keen, H. S., *Theoretical and Experimental Investigation of Microwave Printed Circuits,* Airborne Instrument Laboratory, Inc., November 1956, PB131019, AD110154.

Keen, H. S., *Study of Strip Transmission Lines,* Airborne Instrument Laboratory, Inc., Science Report, December 1, 1955, AD101733.

Kiklen, R., *Stripline Tri-plexer for Use in Narrow-Bandwidth Multi-Channel Filters,* MTT-20, No. 7, July 1972, pp. 486–488.

King, H. E., *Broad-Band Coaxial Choked Coupling Design,* MTT-8, No. 2, March 1960, pp. 132–135.

Klein, G., *Thermal Resistivity Table Simplifies Temperature Calculation,* Microwaves, February 1970, pp. 58–59.

Klopfenstein, R. W., *A Transmission Line Taper of Improved Design,* IRE Proceedings, Vol. 44, No. 1, January 1956, pp. 31–35.

Knechtli, R. C., *Further Analysis of Transission Line Directional Couplers,* IRE Proceedings, Vol. 43, July 1955, pp. 867–869.

Kolker, R. A., *The Amplitude Response of a Coupled Transmission Line, All Pass Network Having Loss,* MTT-15, No. 8, August 1967, pp. 438–443.

Kollberg, E., *Conductor Admittance in Transverse TEM Filters and Slow-Wave Structures,* Electronics Letters, Vol. 3, No. 7, July 1967, pp. 294–296.

Kraker, D. I., *Asymmetric Coupled-Transmission-Line Magic-Tee,* MTT-12, No. 6, November 1967, pp. 595–599.

Kuo, F. F., *Network Analysis and Synthesis,* Second Edition, John Wiley and Sons, New York, 1966.

Kuroda, T., Usui, T., and Yano, K., *Multi-Port Lattice-Type Hybrid Network,* MTT Symposium Digest, 1971, pp. 10–11.

Kurzrok, R. M., *Isolation of Lossy Transmission Line Hybrid Circuits,* MTT-15, No. 2, February 1967, pp. 127–128.

Lagerlof, R. O. E., *End Effects of Half-Wave Stripline Resonators,* MTT-21, No. 5, May 1973, pp. 351–353.

Lange, J., *Interdigitated Stripline Quadrature Hybrid,* MTT-17, No. 12, December 1969, pp. 1150–1151.

LaRosa, R., *Optimum Coupled Resonator Band Pass Filter,* IRE Proceedings, Vol. 47, No. 2, February 1959, p. 329.

Lavendol, L., and Taub, J. J., *Re-entrant Directional Coupler Using Strip Transmission Line,* MTT-13, No. 5, September 1965, pp. 700–701.

Leighton, W. H., and Milnes, A. G., *Junction Reactance and Dimensional Tolerance Effects on X-Band 3-dB Directional Couplers,* MTT-19, No. 10, October 1971, pp. 818–824.

Levine, R. C., *Determination of Thermal Conductance of Dielectric-Filled Strip Transmission Line from Characteristic Impedance,* MTT-15, No. 11, November 1967, pp. 645–646.

Levy, R., *Analysis of Practical Branch-Guide Directional Couplers,* MTT-17, No. 5, May 1969, pp. 289–290.

Levy, R., Directional Couplers, Chapter 3 of *Advances in Microwaves,* Vol. I, Young, L., Editor, Academic Press (New York), 1966.

Levy, R., *General Synthesis of Asymmetric Multi-Element Directional Couplers,* MTT-19, No. 10, October 1971, pp. 818–824.

Levy, R., *New Coaxial to Stripline Transformers Using Rectangular Lines,* MTT-9, No. 3, May 1961, pp. 273–274.

Levy, R., *Tables for Asymmetric Multi-Element Coupled Tranmission Line Directional Couplers,* MTT-12, No. 3, May 1964, pp. 275–279.

Levy, R., *Tables of Element Values for the Distributed Low-Loss Prototype Filter,* MTT-13, No. 5, September 1965, pp. 514–536.

Levy, R., *Transmission-Line Directional Couplers for Very Broadband Operation,* IEE Proceedings (London), Vol. 112, April 1965, pp. 469–476.

Levy, R., and Lind, L. F., *Synthesis of Symmetrical Branch-Guide Directional Couplers,* MTT-16, No. 2, February 1968, pp. 80–89.

Levy, R., and Rozzi, T., *Precise Design of Coaxial Low-Pass Filters,* MTT-16, No. 13, March 1968, pp. 142–147.

Levy, R., and Whiteley, I., *Synthesis of Distributed Elliptic-Function Filters from Lumped-Constant Prototypes,* MTT-14, No. 11, November 1966, pp. 506–517.

Levy, R., and Whiteley, I., *Synthesis of Distributed Elliptic-Function Filters from Lumped-Constant Prototypes,* MTT Symposium Digest, 1966, pp. 83–88.

Lewin, L., *Radiation from Discontinuities in Stripline*, IEE Proceedings (London), Vol. 107, Part C, February 1960, pp. 163–170.

Lind, L. F., *Design of Chebyshev Filters with Shorted Quarter-Wave Stubs and with Uniform Main Line Impedances*, Proceedings of European Microwave Conference, 1969, p. 178.

Lind, L. F., *Synthesis of Asymmetrical Branch-Guide Directional Coupler-Impedance Transformers*, MTT-17, No. 1, January 1969, pp. 45–48.

Lind, L. F., *Synthesis of Equally Terminated Low-Pass Lumped and Distributed Filters of Even Order*, MTT-17, No. 1, January 1969, pp. 43–45.

Luzzatto, G., *On N-Way Hybrid Combiners*, IEEE Proceedings, Vol. 55, No. 3, March 1967, pp. 470–471.

McDermott, M., and Levy, R., *Very Broadband Coaxial DC Returns Derived by Microwave Filter Synthesis*, Microwave Journal, Vol. 8, No. 2, February 1965, pp. 33–36.

McFarland, J. E., and Mohn, R. J., *Exact Analysis of Asymmetric Couplers*, Microwaves, Vol. 2, No. 3, March 1963, pp. 90–93.

McGough, B., *Comment on "Isolation of Lossy Transmission Line Hybrid Circuits"*, MTT-15, No. 11, November 1967, pp. 652–653.

McNaughton, McGarry and Thompson, *A New Broad-Band Coaxial Hybrid Ring*, Report #12,592, General Electric Research Laboratories (England), November 1956.

March, S., *A Wide-Band Stripline Hybrid Ring*, MTT-16, No. 6, June 1968, p. 361.

Marcuvitz, W., *Waveguide Handbook*, MIT Radiation Laboratory Series, Vol. 10, McGraw-Hill (New York), 1957.

Matthaei, G. L., *Comb-Line Bandpass Filters of Narrow or Moderate Bandwidth*, Microwave Journal, Vol. 6, August 1963, pp. 82–91.

Matthaei, G. L., *Design of Wide-Band (and Narrow Band) Bandpass Microwave Filters on the Insertion Loss Basis,* MTT-8, No. 6, November 1960, pp. 580–593.

Matthaei, G. L., *Direct-Coupled Band-Pass Filters with Quarter-Wave Resonators,* IRE National Convention Record, Vol. VI, Part I, 1958, p. 98.

Matthaei, G. L., *Interdigital Band Pass Filters,* MTT-10, No. 6, November 1962, pp. 479–491.

Matthaei, G. L., *Short-Step Chebyshev Impedance Transformers,* MTT-14, No. 8, August 1966, pp. 372–383.

Matthaei, G. L., *Synthesis of Chebyshev Impedance Matching Networks, Filters and Inter-Stages,* Transactions on Circuit Theory, Vol. CT-3, September 1956, pp. 163–172.

Matthaei, G. L., and Cristal, E. G., *Multiplexer Channel-Separating Units Using Interdigital and Parallel-Coupled Filters,* MTT-13, No. 3, March 1965, pp. 328–334.

Matthaei, G. L., Young, L., and Jones, E. M. T., *Microwave Filters, Impedance Matching Networks, and Coupling Structures,* McGraw-Hill (New York), 1964.

Maxwell, S. P., *A Stripline Cavity Resonator for Measurement of Ferrites,* Microwave Journal, Vol. 9, No. 3, March 1966, pp. 99–102.

Metcalf, W. S., *Cascading 4-Port Networks,* Microwave Journal, Vol. 12, No. 9, September 1969, pp. 77–82.

Michelson, M., and Moore, J. F., *Resonator and Preselector in Balanced Strip Line,* MTT-3, No. 2, March 1955, pp. 170–174.

Microwave Associates, *MACAP,* (Microwave Associates Circuit Analysis Program), Not publicly available.

Microwaves Staff, *Product Survey — Stripline Laminates,* Microwaves, January 1968, pp. 101–103.

Millican, G. L., and Wales, R. C., *Practical Strip-Line Microwave Circuit Design,* MTT-17, No. 9, September 1969, pp. 696–705.

Mittra, R., and Itok, T., *Charge and Potential Distributions in Shielded Striplines,* MTT-18, No. 3, March 1970, pp. 149–156.

Mohr, R. J., *New Coaxial Variable Attenuators,* Microwave Journal, Vol. 8, No. 3, March 1965, pp. 99–102.

Mohr, R. J., *Some Design Aspects of Component Utilizing Symmetric 3.0 dB Hybrids,* Microwave Journal, Vol. 5, No. 6, June 1962, pp. 90–94.

Mole, J. H., *Filter Design Data,* John Wiley and Sons (New York), 1952.

Monteath, G. D., *Coupled Transmission Lines As Symmetrical Directional Couplers,* IRE Proceedings, London, England, Part B, Vol. 102, November 1954, pp. 251–268.

Moreno, T., *Microwave Transmission Design Data,* Dover Publications, Inc., (New York), 1958.

Mosko, J. A., *Calculating the Strip Geometry of Offset Parallel-Coupled Strip Transmission Lines Using an IBM 1620 Computer,* Technical Note 4022–66–4, USN Ordnance Test Station, China Lake, California, February 1966.

Mosko, J. A., *Coupling Curves for Offset Parallel Coupled Strip Transmission Lines,* Microwave Journal, Vol. 10, No. 5, April 1967, pp. 35–37.

Mouw, R. B., *Broadband DC Isolator—Monitors,* Microwave Journal, Volume 7, No. 11, November 1964, pp. 75–77.

Mouw, R. B., *A Broad-Band Hybrid Junction and Application to the Star Modulator,* MTT-16, No. 11, November 1968, pp. 911–918.

Mouw, R. B., and Fukuchi, S. M., *Broadband Double Balanced Mixer/Modulators, Part I,* Microwave Journal, Vol. 12, No. 3, March 1969, pp. 131.
Part II, Microwave Journal, Vol. 12, No. 5, May 1969, p. 71.

Muehe, C. E., *Quarter-Wave Compensation of Resonant Discontinuities*, MTT-7, No. 2, April 1959, pp. 296–297.

Nagai, N., *Basic Considerations on TEM-Mode Hybrid Power Dividers*, 1973 MTT Symposium Digest, pp. 218–220.

Nalbandian, V., and Steenaart, W., *Discontinuities in Symmetric Striplines Due to Impedance Steps and Their Compensations*, MTT-20, No. 9, September 1972, pp. 573–577.

Neuf, D., *Generate 1 to 12 GHz Signals with This Double Balanced Frequency Converter*, Microwaves, Vol. 11, No. 3, March 1972, pp. 54–59.

Norgaard, D. E., *The Phase Shift Method of Single Sideband Signal Generation*, IRE Proceedings, Vol. 44, No. 12, December 1956, pp. 1718–1734.

Oliner, A. A., *Effect of Thickness on the Characteristic Impedance of Strip Transmission Line*, Hughes Aircraft Company, July 31, 1953.

Oliner, A. A., *Equivalent Circuits for Discontinuities in Balanced Strip Transmission Lines*, MTT-3, No. 2, March 1955, pp. 134–143.

Oliner, A. A., *The Radiation Conductance of a Series Slot in Strip Transmission Line*, IRE Convention Record 1954, Part 8, p. 78.

Oliver, B. M., *Directional Electro-Magnetic Couplers*, Proceedings IRE, Vol. 42, November 1954, pp. 1686–1692.

Ozeki, H., and Ishii, J., *Synthesis of Stripline Filters*, IRE Transaction on Circuit Theory, Vol. CT-5, June 1958, pp. 104–109.

Packard, K. S., *Machine Methods Make Strip Transmission Lines*, Electronics, 1954, Vol. 27, No. 9, September 1954, pp. 148–150.

Packard, K. S., *Optimum Impedance and Dimensions for Strip-Transmission Lines*, MTT-5, No. 4, October 1957, pp. 244–247.

Parad, L. I., and Moynihan, R. L., *Split-Tee Power Divider*, MTT 13, No. 1, January 1965, pp. 91–95.

Park, D., *Planar Transmission Lines*, MTT-3, No. 3, April 1955, pp. 8–12.

Park, D., *Planar Transmission Lines II,* MTT-3, No. 5, October 1955, pp. 7–10.

Park, D., *Planar Transmission Lines,* MTT-4, No. 2, April 1956, p. 130.

Pascalar, H., *Strip-Line Hybrid Junction,* MTT-5, No. 1, January 1957, pp. 23–30.

Pease, R. L., *Conductor Heating Losses In Strip Transmission Lines with Rectangular Inner Conductors,* USAF-AFCRC Technical Note TN55–395, December 31, 1954.

Pease, R. L., *Power Handling Capacity of Strip Transmission Lines Having Rectangular Inner Conductors with Semi-Circular Rounded Edges,* Research Laboratory, Physical Electronics, Tufts University, Internal Report Mt 8, March 1955, A074083.

Pease, R. L., and Mingins, C. R., *Approximate Universal Formula for Characteristic Impedance of Strip Transmission Lines with Rectangular Inner Conductors,* Research Laboratory, Physical

Electronics, Tufts University, Report 6, September 20, 1954, AO57808.

Pease, R. L., and Mingins, C. R., *A Universal Approximate Formula for Characteristic Impedance of Strip Tranmission Lines with Rectangular Inner Conductors,* MTT-3, No. 2, March 1955, pp. 144–148.

Perini, H., and Sferrazza, P., *Rectangular Waveguide to Strip Transmission Line Directional Coupler,* IRE Wescon Convention Record 1 (1) 1957, pp. 16–21.

Peters, R. W., et al, *Handbook of Tri-Plate Microwave Components,* Sanders Associates, Nashua, New Hampshire, 1956.

Phelan, H. R., *A Wide-Band Parallel-Connected Balun,* MTT-18, No. 5, May 1970, pp. 259–263.

Pon, C. Y., *Control Your Power Split Using a Hybrid Rat-Race,* Microwaves, Vol. 9, No. 4, April 1970, pp. 34–37.

Pon, C. Y., *Hybrid-Ring Directional Couplers for Arbitrary Power Division,* MTT-9, No. 6, November 1961, pp. 529–535.

Pon, C. Y., *A Wide-Band 3 dB Hybrid Using Semi-Circular Coupled Cross-Section,* Microwave Journal, Vol. 12, No. 10, October 1969, pp. 81–85.

Pyle, J. R., *Broad Band Coaxial to Stripline Transitions,* MTT-12, No. 3, May 1964, pp. 364–365.

Pyle, J. R., *Design Curves for Interdigital Band Pass Filters,* MTT-12, No. 5, September 1964, pp. 559–567.

Pyle, J. R., and Green, H. E., *Exact Design of Capacitive Gap Stripline Filters,* Microwaves, May 1966, pp. 28–33.

Rabinowitz, M., and Torgow, E. N., *Design Techniques for Broad Band Strip Line,* Microwave Research Institute, Polytechnic Institute of Brooklyn, March 26, 1956, AD97730.

Ragan, G., *Microwave Transmission Circuits,* MIT Radiation Laboratory Series, Vol. 9, McGraw Hill (New York), 1948.

Reed, J., and Wheeler, G., *A Broadband Fixed Coaxial Power Divider,* IRE National Convention Record, 1957, Vol. V, Part 1, p. 177.

Reed, J., and Wheeler, G., *A Method of Analysis of Symmetrical Four-Port Networks,* MTT-4, No. 4, October 1956, pp. 246–252.

Reich, H. J., Editor, *Very High Frequency Techniques,* Radio Research Lab., Harvard University, 2 volumes, McGraw Hill (New York), 1947.

Reiter, G., and Hammer, G., *Stripline Filters with Coupled $\lambda/8$ Line Sections,* Proceedings of the European Microwave Conference, 1973, Vol. 1, P.B. 3.2.

Ren, C. L., *On the Analysis of General Parallel Coupled TEM Structures Including Nonadjacent Couplings,* MTT-17, No. 5, May 1969, pp. 242–349.

Riblet, H. J., and Reed, J., *Discussion on Synthesis of Narrow-Band Direct Coupled Filters,* IRE Proceedings, Vol. 41, August 1953, pp. 1058–1059.

Riblet, H. J., *Comment on "Synthesis of Symmetrical Branch-Guide Directional Couplers",* MTT-18, No. 1, January 1970, pp. 47–48.

Riblet, H. J., *Synthesis of Narrow-Band Direct Coupled Filters,* IRE Proceedings, Vol. 40, October 1952, pp. 1219–1223.

Riblet, H. J., *General Synthesis of Quarter-Wave Impedance Transformers,* MTT-5, No. 1, January 1957, pp. 36–43.

Richards, P. I., *Resistor-Transmission-Line Circuits,* IRE Proceedings, Vol. 36, February 1948, pp. 217–220.

Richardson, J. K., *An Approximate Formula for Calculating Z_o of a Symmetric Stripline,* MTT-15, No. 2, February 1967, pp. 130–131.

Richardson, J. K., *Gap Spacing for End-Coupled and Side-Coupled Stripline Filters,* MTT-15, No. 6, June 1967, pp. 380–382.

Richardson, J. K., *Gap Spacing for Narrow-Bandwidth End-Coupled Symmetric Stripline Filters,* MTT-16, No. 8, August 1968, pp. 559–560.

Richardson, J. K., *Graphical Design of Stripline Directional Couplers,* Microwaves, October 1967, pp. 71–74.

Richardson, J. K., *Precise Strip-Line Filter Design Eliminates Resonator Trimming,* Microwaves, Vol. 8, No. 7, July 1969, pp. 52–54.

Robinson, L. A., *Wideband Interdigital Filters with Capacitively Loaded Resonators,* MTT Symposium Digest, 1965, pp. 33–37.

Robinson, S. J., *Broadband Hybrid Junctions,* MTT-8, No. 6, November 1960, pp. 671–672.

Rosenzweig, A., *Determining the Characteristic Impedance of Non-Symmetrical Strip Transmission Line,* LEL Div. of Varian, Technical Memo TM 0002, April 1967.

Roy, S. C. D., *On High-Pass Transmission-Line Directional Couplers,* MTT-17, No. 7, July 1969, pp. 400–401.

Rustogi, O. P., *Linearly Tapered Transmission Line and Its Application to Microwaves,* MTT-17, No. 3, March 1969, pp. 166–168.

Rutz, E., *A Stripline Frequency Translator,* MTT-9, No. 2, March 1961, pp. 158–161.

Saad, T. S., Editor, *Microwave Engineers Handbook,* Volume I and Volume II, Artech House (Dedham, Mass.), 1971.

Saal, R., *Der Entwurf Von Filtern mit Hilfe des Kataloges Normierten Tiefpasse,* Telefunken GMBH (Backnang, W. Germany), 1963.

Saal, R., and Ulbrich, *On The Design of Filters by Synthesis,* IRE Transactions on Circuit Theory, December 1958, pp. 284–327.

Sato, R., and Cristal, E. G., *Simplified Analysis of Coupled Transmission-Line Networks,* MTT-18, No. 3, March 1970, pp. 122–130.

Schelpenoff, S. A., *Electromagnetic Waves,* Van Nostrand (New York), 1943.

Schetzen, M., *Printed Microwave Systems,* MIT Technical Report 289, September 30, 1954, pp. 1–37, AD66376.

Schiffman, B. M., *Capacitively-Coupled Stub Filter,* MTT-13, No. 2, March 1965, pp. 253–254.

Schiffman, B. M., *A Harmonic Rejection Filter Designed by an Exact Method,* MTT-12, No. 1, January 1964, pp. 58–60.

Schiffman, B. M., *A New Class of Broadband Microwave 90° Phase Shifters,* MTT-6, No. 2, April 1958, pp. 232–237.

Schiffman, B. M., *Realizations of a Duo-Pole Branch of an Elliptic-Function Bandstop Filter,* MTT-5, No. 8, August 1967, p. 487.

Schiffman, B. M., *Two Nomograms for Coupled Line Sections for Bandstop Filters,* MTT-14, No. 6, June 1966, pp. 297–299.

Schiffman, B. M., and Matthaei, G. L., *Exact Design of Band-stop Microwave Filters,* MTT-12, No. 1, January 1964, pp. 6–15.
Cristal, E. G., *Addendum to "An Exact Method for Synthesis of Microwave Band-Stop Filters" (sic),* MTT-12, No. 3, May 1964, pp. 369–382.

Schiffres, P., *How Much CW Power Can Striplines Handle?,* Microwaves, June 1966, pp. 25–34.

Schumacher, H. L., *Dissipation Loss of Chebyshev Bandpass Filters,* Microwave Journal, Vol. 10, No. 9, August 1967, pp. 41–43.

Sciegenny, J., and Schetzen, M., *Strip Transmission Systems,* Quarterly Progress Report of the Research Laboratory of Electronics, MIT, Part XI, 1953, p. 83.

Scott, H. J., *The Hyperbolic Transmission Line as a Matching Section,* IRE Proceedings, Vol. 41, November 1953, pp. 1654–1657.

Sferrazza, P. J., *Unidirectional Coupler Research and Development Program,* Final Report No. 5224–1240–8, Sperry Gyroscope Co., New York, 1953.

Sharpe, C. B., *An Equivalence Principle for Non-Uniform Transmission Line Directional Couplers,* MTT-15, No. 7, July 1967, pp. 398–405.

Shelton, J. P., *Impedances of Offset Parallel Coupled Strip Transmission Lines,* MTT-14, No. 1, January 1966, pp. 7–15.
Correction to "Impedances of Offset Parallel Coupled Strip Transmission Lines", MTT-14, No. 5, May 1966, p. 249.

Shelton, J. P., *Multiple Line Directional Couplers,* IRE National Convention Record, 1957, Vol. V, Part 1, p. 254.

Shelton, J. P., and Mosko, J., *Design Tables for Directional Couplers and Phase Shifters,* Library of Congress Photoduplication Service, Doc. # 9017.

Shelton, J. P., and Mosko, J. A., *Synthesis and Design of Wide-Band Equal-Ripple TEM Directional Couplers and Fixed Phase Shifters,* MTT-14, No. 10, October 1966, pp. 462–473.

Shelton, J. P., Von Wagoner, R., and Wolfe, J., *Tandem Couplers and Phase Shifters: A New Class of Unlimited Bandwidth Components,* 14th Annual Symposium, USAF, Wright-Patterson AFB, October 1964.

Shelton, J. P., Wolfe, J., and Von Wagoner, R., *Tandem Couplers and Phase Shifters for Multi-Octave Bandwidth,* Microwaves, April 1965, pp. 14–19.

Shimizu, J. K., *A Stripline Directional Coupler,* Scientific Report 1, Project 1592, Contract AF19 (604) –1571, AD117286, Stanford Research Institute, June 1957.

Shimizu, J. K., *Stripline 3-dB Directional Couplers,* IRE Wescon Convention Proceedings, (1) 1, 1957, pp. 4–15.

Shimizu, J. K., and Jones, E. M. T., *Coupled-Transmission-Line Directional Couplers,* MTT-6, No. 4, October 1958, pp. 403–410.

Singletary, J. J., *Fringing Capacitance in Stripline Coupler Design,* MTT-14, No. 8, August 1966, p. 398.
Corrections to "Fringing Capacitance in Stripline Coupler Design", MTT-15, No. 3, March 1967, p. 200.

Sisodia, M. L., and Gandhi, O. P., *Octave Bandwidth L- and S-band Stripline Discriminators,* MTT-15, No. 4, April 1967, pp. 271–272.

Slater, J. C., *Microwave Transmission,* McGraw-Hill, New York, 1942.

Sleven, R. L., *Save Space with Stripline Design,* Microwaves, Vol. 9, No. 9, September 1970, pp. 71–74.

Smith, H., *Computer-Generated Tables for Filter Design,* Electronic Design, May 10, 1963, pp. 54–57.

Sobol, H., *A Microwave Hybrid with Impedance Transforming Properties,* MTT-19, No. 9, September 1971, pp. 774–776.

Sommers, D., *Slot Array Employing Photo-Etched Tri-Plate Transmission Lines,* MTT-3, No. 2, March 1955, pp. 157–162.

Southworth, G. C., *Principles and Applications of Waveguide Transmission,* Van Nostrand, New York, 1950.

Spector, N., *Evaluation of Power Capacity of Stripline,* Proceedings of the National Electrical Conference, October 1956, pp. 715–722.

Standley, R. D., and Todd, A. C., *Discontinuity Effects in Single Resonator Traveling Wave Filters,* MTT-11, No. 6, November 1963, pp. 551–552.

Standley, R. D., and Todd, A. C., *A Note on Stripline Bandstop Filters with Narrow Stop Bands,* MTT-11, No. 6, November 1963, pp. 548–549.

Stearns, W. P., *Advances in Miniaturized Microwave Solid-State Receivers,* Electronic Engineer, November 1966, pp. 66–73.

Steenaart, W. J. D., *The Synthesis of Coupled Transmission Line All-Pass Networks in Cascades of 1 to n,* MTT-11, No. 1, January 1963, pp. 23–29.

Steenaart, W. J. D., *A Note on the Synthesis of Coupled-Line Directional Couplers and All-Pass Networks,* MTT-16, No. 8, August 1968, pp. 554–555.

Stinehelfer, H. E., *Microstrip Circuit Design,* Microwave Journal, Vol. 13, No. 5, May 1970, pp. 71–74.

Stinehelfer, H. E., *Microstrip Circuit Designs,* Technical Report AFAL-TR-69–10, Air Force Avionics Laboratory, Wright-Patterson AFB, February 1969.

Tatsuguchi, I., *UHF Strip Transmission Line Hybrid Junction,* MTT-9, No. 1, January 1961, pp. 3–6.

Taub, J. J., and Kurpis, G. P., *A More General N-Way Hybrid Power Divider,* MTT-17, No. 7, July 1969, pp. 406–408.

Taub, J. J., *When to Use Strip Transmission Line,* Electronic Design, September 27, 1961, p. 184.

Tell, P., *A Brief History of Microwave Printed Circuits with Bibliography,* Report B1–2, Tellite Corp. (Orange, New Jersey), October 9, 1962.

Tenenholtz, R., *Broadband MIC Multithrow PIN Diode Switches,* Microwave Journal, Vol. 16, No. 7, July 1973, pp. 25–30.

Tetarenko, R. P., and Goud, P. A., *Broad-Band Properties of a Class of TEM-Mode Hybrids,* MTT-19, No. 11, November 1971, pp. 887–889.

Thompson, W. E., *N-Way Hybrid Combiner,* Electronics Letters, Vol. 3, No. 11, November 1967, pp. 584–586.

Tomizasu, K., *Attenuation in a Resonant Ring Circuit,* MTT-8, No. 2, April 1960, pp. 253–254.

Torgow, E. N., *Microwave Filters,* Electro-Technology, Vol. 67, April 1961, pp. 90–96.

Torgow, E. N., *Microwave Stripline Program,* Microwave Research Institute, Polytechnic Institute of Brooklyn, Final Report R-476–56, PIB-406, March 1956.

Torgow, E. N., and Griesmann, J. W. E., *Miniature Strip Transmission Line for Microwave Application,* Microwave Research Institute, Polytechnic Institute of Brooklyn, February 4, 1955, AD68546.

Torgow, E. N., and Griesmann, J. W. E., *Miniature Strip Transmission Line for Microwave Applications,* MTT-3, No. 2, March 1955, pp. 57–64.

Torgow, E. N., and Griesmann, J. W. E., *Stripline,* Polytechnic Institute of Brooklyn, Report R-360–54, PIB 294, February 1954.

Toulios, P. P., and Todd, A. C., *Synthesis of Symmetrical TEM-Mode Directional Couplers,* MTT-13, No. 5, September 1965, pp. 536–544.

Tresselt, C. P., *Broad-Band High IF Mixers Based on Magic T's,* MTT-18, No. 1., January 1970, pp. 58–60.

Tresselt, C. P., *Broadband Tapered-Line Phase Shift Networks,* MTT-16, No. 1, January 1968, pp. 51–52.

Tresselt, C. P., *Design and Computed Theoretical Performance of Three Classes of Equal-Ripple Nonuniform Line Couplers,* MTT-17, No. 4, April 1969, pp. 218–230.

Tresselt, C. P., *The Design and Construction of Broadband, High-Directivity, 90° Couplers Using Non-Uniform Line Techniques,* MTT-14, No. 12, December 1966, pp. 647–656.

Tyminski, W. V., and Hyles, A. E., *A Wide-Band Hybrid Ring for UHF,* IRE Proceedings, Vol. 41, January 1953, pp. 81–87.

Tyrell, W. A., *Hybrid Circuits for Microwaves,* IRE Proceedings, Vol. 35, 1947, pp. 1307–1313.

Udelson, B. J., and McDonald, D. W., *The Design of Two-Section Transformers Having Variable Bandwidth,* Microwave Journal, Vol. 9, No. 4, April 1966, pp. 68–70.

Van Dover, L. K., *Round-Robin Correlation Testing of an X-band Dielectric Constant and Dissipation Factor Test Method,* Proceedings of the 11th Electrical Insulation Conference, 1973, pp. 269–270.

Van Patten, R. A., *Design of Improved Microwave Low-Pass Filters Using Stripline Techniques,* IRE National Convention Record (1) 5, 1957, pp. 197–207.

Van Patten, R. A., *Design of Microwave Low-Pass Filters Using Stripline Techniques,* Rome Air Development Center, October 1956, AD97719.

Van Wagoner, R., *Multi-Octave Bandwidth Microwave Mixer Circuits,* MTT Symposium Digest, 1968, pp. 8–15.

Van Wagoner, R., and Shelton, P., *Monopulse Comparator Networks for Multi-Octave Operation,* MTT Symposium Digest, 1965, pp. 187–192.

Vendelin, G. D., *Limitations on Stripline Q,* Microwave Journal, Vol. 13, No. 5, May 1970, pp. 63–69.

Von Hippel, A. R., *Dielectrics and Waves,* John Wiley and Sons (New York), 1954.

Von Hippel, A. R., *Dielectric Materials and Applications,* John Wiley and Sons (New York), 1954.

Vossberg, W., *Stripping the Mystery from Stripline Laminates,* Microwaves, January 1968, pp. 104–112.

Waldron, R. A., *Theory of the Stripline Cavity Resonator,* Marconi Review 152, Vol. 27, 1st Quarter, 1964, pp. 30–42.

Walker, S., *Broadband Stripline Balun Using Quadrature Couplers,* MTT-16, No. 12, February 1968, pp. 132–133.

Wanselow, R. D., and Tuttle, L. P., Jr., *Practical Design of Strip Transmission Line Half Wavelength Resonator Directional Filters,* MTT-7, No. 1, January 1959, pp. 168–173.

Wells, R. E., *Microwave Bandpass Filter Design Part 1* Microwave Journal, Vol. 5, November 1962, pp. 92–98.
Part 2, December 1962, pp. 82–88.

Wenzel, R. J., *Exact Design of TEM Microwave Networks Using Quarter-Wave Lines,* MTT-12, No. 1, January 1964, pp. 94–111.

Wenzel, R. J., *Printed Circuit Complimentary Filters for Narrow Bandwidth Multiplexers,* MTT-16, No. 3, March 1968, pp. 147–157.

Wenzel, R. J., *Theoretical and Practical Applications of Capacitance Matrix Transformations to TEM Network Design,* MTT Symposium Digest, 1966, pp. 94–99.

Wheeler, H. A., *Transmission Line Properties of Parallel Strips Separated by a Dielectric Sheet,* MTT-13, No. 2, March 1965, pp. 172–185.

Wheeler, H. A., *Transmission Line Properties of Parallel Wide Strips by a Conformal-Mapping Approximation,* MTT-12, No. 3, May 1964, pp. 280–289.

Wheeler, H. A., *Transmission Lines with Exponential Tapers,* IRE Proceedings, Vol. 27, January 1939, pp. 65–71.

White, D. R. J., *Developments in Printed Microwave Components,* Electronic Industries and Tele-Tech, 1957, Vol. 16, No. 11, p. 63.

Wild, N. R., *Photo-Etched Microwave Transmission Lines,* Electronic Industries and Tele-Tech, February and March 1955.

Wild, N. R., *Photo-Etched Microwave Transmission Lines,* MTT-3, No. 2, March 1955, pp. 21–30.

Wilds, R. B., *Microwave Two-Phase Converters for Imageless Receivers,* Microwave Journal, Vol. 4, No. 9, September 1961, pp. 84–87.

Wilkinson, E., *An N-Way Hybrid Power Divider,* MTT-8, No. 1, January 1960, pp.116–118.

Wilkinson, E., and LaRussa, F., *Contraphase Shifter,* MTT-10, No. 5, September 1962, pp. 390–391.

Willis, J., and Sinha, N. K., *Non-Uniform Transmission Lines As Impedance Transformers,* IRE Proceedings, Vol. 103B, March 1956, pp. 166–172.

Womack, C. P., *The Use of Exponential Transmission Lines in Microwave Components,* MTT-10, No. 2, March 1962, pp. 124–132.

Worboys, D. C., *Selecting Microwave Stripline Laminates,* Proceedings of the Eleventh Electrical Insulation Conference, 1973, pp. 258–260.

Yamamoto, S., Azakami, T., and Itakura, K., *Coupled Strip Transmission Line with Three Center Conductors,* MTT-14, No. 10, October 1966, pp. 446–461.

Yamamoto, S., Azakami, T., and Itakura, K., *Slit-Coupled Strip Transmission Lines,* MTT-14, No. 11, November 1966, pp. 543–553.

Yamamoto, S., Azakami, T., and Itakura, K., *Coupled Non-Uniform Transmission Line and Its Application,* MTT-15, No. 4, April 1967, pp. 220–231.

Yamashita, E., and Atsuki, K., *Stripline with Rectangular Outer Conductor and Three Dielectric Layers,* MTT-18, No. 5, May 1970, pp. 238–243.

Yamashita, E., and Yamayaki, S., *Parallel-Strip Line Embedded in or Printed on a Dielectric Sheet,* MTT-16, No. 11, November 1968, pp. 972–973.

Yee, H., Chang, F., and Audek, H., *A Study of Wideband N-Way Power Dividers,* Final Technical Report, University of Alabama, Research Institute (Huntsville, Alabama), AD693290, July 1969.

Yee, H., Chang, F., and Audek, H., *N-Way TEM-Mode Broadband Power Dividers,* MTT-18, No. 10, October 1970, pp. 682–688.

Young, L., *The Analytical Equivalence of TEM-mode Directional Couplers and Transmission Line Stepped Impedance Filters,* IEE Proceedings (London), Vol. 110, February 1963, pp. 275–281.

Young, L., *Coaxial Stub Design,* Electronics, Vol. 30, July 1957, p. 188.

Young, L., *Design Chart for Quarter-Wave Stubs,* Microwave Journal, Vol. 4, No. 5, May 1961, p. 92.

Young, L., *Direct-Coupled Cavity Filters for Wide and Narrow Bandwidths,* MTT-11, No. 3, May 1963, pp. 162–178.

Young, L., *Inhomogeneous Quarter-Wave Transformers,* Microwave Journal, Vol. 5, No. 2, February 1962, pp. 84–89.

Young, L., *Microwave Filters – 1965,* MTT-13, No. 5, September 1965, pp.489–508.

Young, L., *Microwave Filters Using Parallel Coupled Lines (Reprint Volume)*, Artech House (Dedham, Mass.), 1972.

Young, L., *Optimum Quarter-Wave Transformers*, MTT-8, No. 5, September 1960, pp. 478–482.

Young, L., *Parallel Coupled Lines and Directional Couplers, (Reprint Volume)*, Artech House (Dedham, Mass.), 1972.

Young, L., *The Practical Realization of Series-Capacitive Couplings for Microwave Filters*, Microwave Journal, Vol. 5, No. 12, December 1962, pp. 79–81.

Young, L., *Progress in Circuit Theory, 1960–1963: Part 3 on Microwave Filters*, CT-11, March 1964, pp. 10–12.

Young, L., *The Quarter-Wave Transformer Prototype Circuit*, MTT-8, No. 5, September 1960, pp. 483–489.

Young, L.., *Stepped-Impedance Transformers and Filter Prototypes*, MTT-10, No. 5, September 1962, pp. 339–359.

Young, L., *Synchronous Branch Guide Directional Couplers for Low and High Power Applications*, MTT-10, No. 6, November 1962, pp. 457–475.

Young, L., *Tables for Cascaded Homogeneous Quarter-Wave Transformers*, MTT-7, No. 2, April 1959, pp. 233–237.

Young, L., Matthaei, G., and Jones, E. M. T., *Microwave Bandstop Filters with Narrow Stop Bands*, MTT-10, No. 6, November 1962, pp. 416–427.

Zverev, A. I., *Handbook of Filter Synthesis*, John Wiley and Sons, New York, 1967.

Index